*Front Cover;* The northern Malverns from below Herefordshire Beacon

**Plate I** (*Frontispiece*)

Aerial view of the
Ludlow Anticline
looking north-eastwards
from near Leinthall
Earls
(*Cambridge University
Aerial Photograph
No. ACY 69*)

(*For full explanation see p. ix*)

NATURAL ENVIRONMENT RESEARCH COUNCIL
INSTITUTE OF GEOLOGICAL SCIENCES

British Regional Geology

# The Welsh Borderland

(THIRD EDITION)

By J. R. Earp, M.Sc., Ph.D. and
B. A. Hains, B.Sc., Ph.D.

*Based on previous editions by*

R. W. Pocock, D.Sc. and T. H. Whitehead, M.Sc.

LONDON
HER MAJESTY'S STATIONERY OFFICE
1971

*The Institute of Geological Sciences
was formed by the incorporation of the
Geological Survey of Great Britain
and the Museum of Practical Geology
with Overseas Geological Surveys
and is a constituent body of the
Natural Environment Research Council*

# Foreword to Third Edition

The content of former editions of this compact digest of Welsh Borderland geology, the second edition of which ran to 10 impressions, has remained substantially unchanged since the first edition was published in 1935. Since then almost every geological formation within the region has been the subject of further research so that a wealth of new literature is now available.

This edition attempts to take account of all significant work done in the region during the past 35 years, while retaining the layout and scope of the original publication. Chapters 2, 3, 4 and parts of 9 and 11 have been prepared by Dr. Hains and the remainder by Dr. Earp who has also edited the work. It is hoped that the somewhat greater attention to detail and the use of a more modern scientific vocabulary has not impaired the value of the work as a semi-popular guide to the geology of this fascinating region which has nurtured and stimulated so many generations of field geologists.

The authors' grateful acknowledgments are due to:

The University of Cambridge Committee for Aerial Photography for permission to reproduce Plates I, II and XA.

The Council of the Geological Society of London for permission to illustrate data shown here as Figs. 5, 10, 11, 17, 18, 19, 20, 33, 34, 37 and 40 B, C.

The Council of the Geologists' Association for permission to illustrate data shown here as Figs. 25 and 26.

The Trustees of the British Museum (Natural History) for permission to illustrate data shown here as Figs. 23 D, E, 32 B, N, O, 36 E, F, H, 40 G, J, K, L, and 41 A, B; also Plate VII.

The Editors and Publishers of the Geological Magazine for permission to illustrate data shown here as Figs. 27 and 42.

The Editors and Publishers of the Liverpool and Manchester Geological Journal for permission to illustrate data shown here as Fig. 35.

Our indebtedness to the authors named in the captions to the above illustrations is also gratefully acknowledged.

Institute of Geological Sciences,
Exhibition Road,
South Kensington,
London, S.W.7.
4th January, 1971

K. C. Dunham
*Director*

*An EXHIBIT illustrating the geology and scenery of the region described in this handbook is set out in the Museum of Practical Geology, Institute of Geological Sciences, Exhibition Road, South Kensington, London, S.W.7.*

# Contents

# Illustrations

## Figures in text

# Plates

---

[1]Numbers preceded by A refer to photographs in the Geological Survey collection.

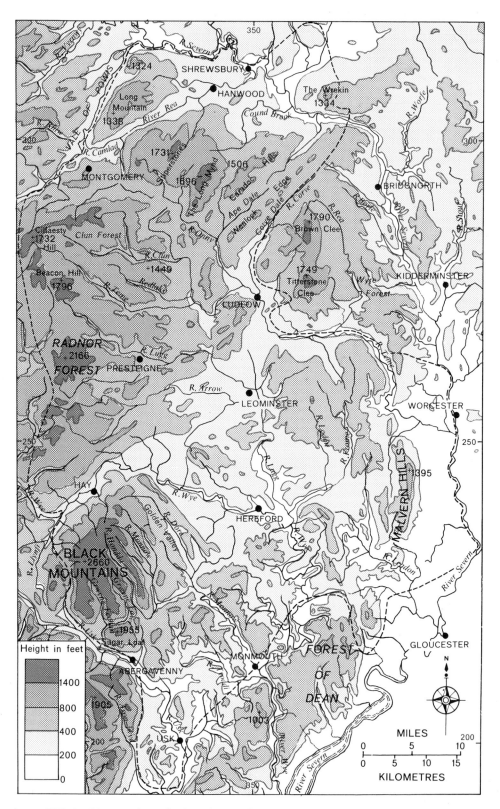

FIG. 1. *Map to show the boundary and main physical features of the region.*

# 1. Introduction

In its wider sense the Welsh Borderland is the area adjacent to the boundary between England and Wales from Chester in the north to Newport in the south. The region covered by this account (Fig. 1) excludes the Marches north of the latitude of Welshpool and Shrewsbury and in the south it includes only part of Monmouthshire. It takes in a portion of the county of Worcestershire east of the Malvern and Abberley Hills which would not normally be considered to be Welsh Borderland country at all.

## History of Research

Since the earliest days of the history of the science of geology the Welsh Borderland has attracted the attention of geologists by the great variety and interest of its formations; for in no other area perhaps can the sequence of the Palaeozoic rocks be seen to such advantage and within such a comparatively small district.

The first comprehensive investigation of the rocks of the area was made by Sir Roderick Murchison, who, in 1835, introduced the well-known name 'Silurian' for the series of rocks which he had studied in the land of the old British tribe of the Silures (the southern Marches); he divided this system into an upper and a lower series.

In the meantime Professor Sedgwick had worked out the succession of the Palaeozoic rocks of North Wales and, in 1835, proposed the name Cambrian for this sequence, adopting Murchison's name Silurian for the overlying rocks in the Berwyn Mountains. It was then found that the lower part of the Silurian of Murchison and the upper part of the Cambrian of Sedgwick were in part equivalent, and in 1879 Professor Lapworth suggested the name 'Ordovician' (from the tribe of the Ordovices which inhabited North Wales) for the middle portion of the Cambro-Silurian sequence, the designation of which was in dispute. Lapworth's classification of the older Palaeozoic rocks into Cambrian, Ordovician and Silurian is now accepted.

Murchison's great work *The Silurian System*, published in 1839, still remains a fund of information and a basis for the work of investigators in the region and, with his *Siluria* published in 1854 (and four later editions), forms a monument to the knowledge and industry of this pioneer in the science, whose researches embraced not only the older Palaeozoic rocks but also the earlier Pre-Cambrian and the later Old Red Sandstone and Carboniferous rocks of the district.

Among other early workers whose well-known names may be mentioned are Aitkin, Lewis, Prestwich, Phillips, Salter, Aveline, Lightbody, Allport, Bonney, Maw, Morton, Callaway and Blake, who have all contributed to the elucidation of the geology of this complex region. We may also refer to W. S. Symonds, who, in his *Records of the Rocks*, 1872, dealt in a charming and interesting manner with the geology, natural history and antiquities, and to J. D. La Touche, whose *Handbook to the Geology of Shropshire*, 1884, was a useful guide to the more important localities and contained numerous drawings of typical fossils.

The original maps of the Geological Survey on the scale of one inch to one mile were published between the years 1844 and 1855, and there were revisions up to 1873. A number of horizontal sections were also produced on a scale of six inches to one mile.

In the later nineteenth century and early twentieth, research was stimulated by Charles Lapworth and W. W. Watts; the former was the discoverer of the '*Olenellus*' fauna in the Lower Cambrian rocks of Shropshire and worked on the Ordovician rocks of the Shelve district, while the latter improved our knowledge of the geology of the Breidden Hills, and other areas.

Important work was also done by W. S. Boulton on the Uriconian rocks, by E. S. Cobbold, C. J. Stubblefield and O. M. B. Bulman on the Cambrian, by G. L. Elles, I. L. Slater, E. M. R. Wood, C. I. Gardiner, W. F. Whittard, B. B. Bancroft and W. T. Dean on the Ordovician and Silurian and by W. Wickham King on the Old Red Sandstone.

In the structurally complex Malvern and Abberley areas the work of John Phillips, Charles Callaway and others was followed by that of T. T. Groom whose papers were published around 1900. At least six successive attempts to improve upon Groom's interpretation of the structure of this line of disturbance have been made since 1947. In Radnorshire small inliers along the structurally complex line of the Church Stretton Fault have been investigated by J. E. Davis, T. C. Cantrill, A. H. Cox and, the largest at Old Radnor, by E. J. Garwood and E. Goodyear.

Since the first World War the Geological Survey has mapped on the six-inch scale and published explanatory memoirs on the areas covered by the Shrewsbury (152), Church Stretton (166) and Droitwich (182) sheets, and, in the extreme south of the region, there are recent six-inch maps and memoirs relating to small areas which fall within the Monmouth (233), Newport (249) and Chepstow (250) sheets.

Following the work of S. H. Straw on the Ludlow succession at Builth just outside this region, the Silurian rocks, especially those of the Ludlow Series, have been extensively investigated. From 1952 onwards this work was much stimulated by collaboration among a group of geologists who formed a '*Ludlow Research Group*', its activities co-ordinated for many years by J. D. Lawson and V. G. Walmsley. The Old Red Sandstone has also been the subject of many recent papers, especially by members of the Geological Department of the Natural History Museum on its fish faunas and early terrestrial plants. Sedimentological studies on almost every formation in the Welsh Borderland have tended to swell the volume of literature relating to the region in the past decade.

## Physical Features and Drainage

The main physical features of the Welsh Borderland may best be described from north to south.

The district is bounded on the north by the great plain of north Shropshire, floored by Coal Measures and Triassic rocks which rest upon the northern slopes and spurs of the old Palaeozoic mass to the south and are largely obscured by a mantle of drift deposits, consisting of boulder-clay and glacial sands and gravels, brought into the district by ice both from the Irish Sea and from the Welsh Mountain area to the west.

The River Severn, descending from the high ground of Wales, enters upon the Shropshire plain a few miles below Welshpool, meanders eastward along its southern border, cutting, in places, across spurs of the older rocks that project into it, and finally leaves the plain by the narrow gap of the Iron-bridge Gorge to flow southwards to the Bristol Channel.

The first important feature met with south of the river is the striking mass of the Breidden rising to 1202 ft (366 m), with Moel-y-Golfa, 1324 ft (403 m), the former a remarkable laccolite of dolerite the general structure of which is anticlinal, and the latter an intrusion of andesite.

To the south of the Breidden mass lies the syncline of the Long Mountain composed of Upper Silurian rocks rising to a height of 1338 ft (408 m).

The Shelve district of Ordovician rocks, rising to a culminating point of 1684 ft (513 m) in the great laccolite of dolerite known as the Corndon, is separated from the Long Mountain syncline by a broad valley cut in soft Silurian shales and drained by small streams flowing in opposite directions; actually at times from the same pool (Marton Pool) on the low watershed. The Shelve Ordovician area is bounded on the south-east by the impressive ridge of the Stiperstones, formed of Arenig quartzite; the jagged crags of this resistant rock, piercing the skyline, can be seen from a great distance.

A valley cut in Cambrian shales separates the Shelve Ordovician mass from an area of Pre-Cambrian rocks including the plateau of the Long Mynd, which is composed of an immense thickness of Pre-Cambrian grits, flags and conglomerates with high to vertical dip. This plateau, 4 to 5 square miles (10 to 13 sq. km) in area, rises to a maximum height of 1696 ft (517 m) and, though approximately level, slopes gradually to the north and south. Its steep edges are deeply trenched by streams forming the valleys known locally as 'batches' or 'gutters'.

A spur of the Longmyndian rocks extends north-eastwards in Bayston and Sharpstone Hills and, beyond the River Severn, rises again in the outstanding mass of Haughmond Hill.

The Long Mynd is flanked on either side by outcrops of Pre-Cambrian volcanic rock—the Western and Eastern Uriconian groups. The Western Uriconian gives rise to the prominent mass of Pontesford Hill and smaller igneous areas to the south-west. The Eastern Uriconian is developed along the great Church Stretton Fault, in the remarkable range of hog-backed hills, Ragleth, Caradoc, the Lawley and, in continuation of their line, those of the Wrekin, Ercall and Lilleshall to the north of the River Severn.

Ordovician and Silurian rocks form a succession of ridges and valleys south-east of, and parallel to, the Eastern Uriconian range. The lowest beds of the Ordovician are the grits which give rise to the scarp of Hoar Edge. The Wenlock Limestone forms the very regular lower scarp of Wenlock Edge, about 16 miles (25 km) long, and the Aymestry Limestone a parallel but more dissected scarp along which the best known eminences are Callow Hill, Norton Camp and View Edge.

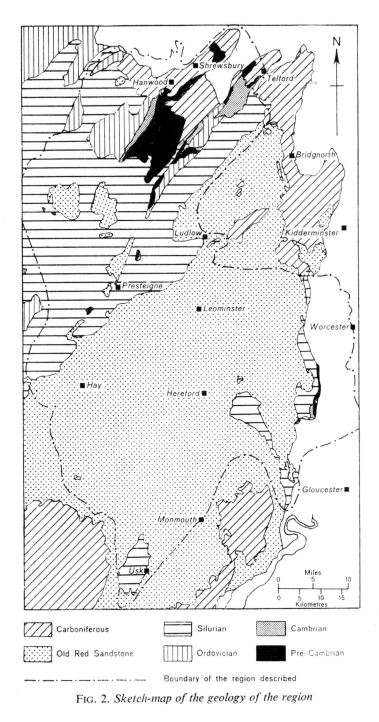

FIG. 2. *Sketch-map of the geology of the region*

To the south-east of the Aymestry Limestone ridge and separated from it by the broad valley of Corve Dale lies a triangular plateau formed of the higher beds of the Lower Old Red Sandstone. Two table-topped masses rise from this plateau, the Brown Clee, 1792 ft (546 m), and Titterstone Clee, 1749 ft (533 m) (see British Regional Geology, Central England). These are outlying relics of Coal Measures which have been protected from denudation by thick coverings of contemporaneous dolerite or basalt.

The southern slopes of the Long Mynd and Shelve country are drained by the rivers Onny and Camlad, which also drain the northern slopes of the plateau of Clun Forest—an area of Ludlovian and Downtonian rocks rising to a height of 1796 ft (547 m) in Beacon Hill.

The central part of the Clun Forest district is drained by the rivers Clun and Teme; the latter, flowing by Downton Castle, Ludlow and Tenbury, forms the southern boundary of the tracts of high ground mentioned above, most of which lie within the county of Shropshire.

Ludlow is situated on the north-eastern rim of a plunging anticline of Silurian rocks in the core of which lies the Vale of Wigmore on the Wenlock Shale. In this area the wooded ridges formed by the Wenlock and Aymestry limestones are well-marked scenic features.

South of Ludlow, around Leominster, Hereford and Monmouth, an extensive area, mainly of Old Red Sandstone, presents a landscape of rather subdued hills or rolling terrain traversed by the broad plain of the River Wye and by the valleys of its tributaries the Lugg and the Monnow. On the west of this area the continuation south-westwards of the Church Stretton line of faulting brings to the surface the interesting area of Old Radnor, where Silurian strata can be seen resting unconformably on Pre-Cambrian (Longmyndian) rocks. Here also are the igneous masses of Hanter, Worsel Wood and Stanner, the age of which is probably Pre-Cambrian, and nearby is the prominent mass of Radnor Forest, rising to a height of 2166 ft (660 m) and formed of siltstones of the Ludlow Series. Farther south the most striking feature of the landscape is the great plateau of the Black Mountains, composed of near-horizontal Old Red Sandstone, which forms a bold scarp overlooking the Wye Valley and rises to a height of 2660 ft (811 m) at Waun Fach and to 2624 ft (799 m) at Pen y Gader Fawr. South-east of the main scarp the plateau is deeply dissected by southwardly flowing streams such as the Afon Honddu, which joins the Monnow, and the Grwyne Fawr which joins the Usk.

The region includes in the east the Silurian, Cambrian and Pre-Cambrian rocks of the Malvern range, beyond which lies the Triassic plain of Worcester. The summit of North Hill, Malvern, 1307 ft (398 m), is the highest point of the range.

The Silurian inlier of May Hill lies on the continuation southward of the Malvern axis and, with the Silurian district of Ledbury and the Silurian inlier of Woolhope, shows the same type of scenery (due to alternation of wooded limestone scarps with valleys cut in the intervening soft shales) as that in the Ludlow and Wenlock country.

The Old Red Sandstone tract continues southward between Abergavenny and Monmouth to the Silurian inlier of Usk flanked by the Carboniferous

Limestone outcrops of the South Wales syncline on the west and of the
Forest of Dean syncline on the east.


## Geological History

Traces of the remote Pre-Cambrian history of the region are preserved in
the metamorphic rocks of Rushton, Primrose Hill and the Malvern Hills,
and the prolonged and complex events which produced these highly meta-
morphosed rocks may have ended considerably before 1200 million years
ago. This ancient phase of metamorphism was followed in early Proterozoic
times by a major outburst of volcanic activity during which mainly andesitic
and rhyolitic lavas were poured out and much volcanic ash was ejected
along lines of crustal weakness trending broadly from north to south. Small
scale intrusive activity also occurred. After an interval of unknown duration,
following the vulcanism and spanning a long period some time between
1200 million and 600 million years ago, sedimentation occurred in a fairly
narrow crustal depression the margins of which may have been controlled
by lines of weakness approximating to the present disturbances of Church
Stretton and Pontesford–Linley. During this phase something of the order
of 25 000 ft (7620 m) of sediment was accumulated with only one significant
break in continuity. The immense panorama of Pre-Cambrian history was
completed by a phase of earth movements followed by a long period of
erosion.

The Cambrian period was heralded by a great transgression of the sea. The
Pre-Cambrian formations had been strongly folded and denuded to an almost
level surface before the oldest Cambrian rocks were laid down. The type of
sedimentation met with in the Lower Cambrian suggests that the sea was
then shallow and subject to current and wave action. Earth movements took
place during Cambrian time, as proved by unconformities within the system,
while the presence of a great thickness of uniformly fine sediment in the
Upper Cambrian seems to indicate a progressive lowering of the sea floor.

At the close of the Cambrian period a shallowing of the sea took place
not only by the accumulation of sediment but also by an elevation of the sea
floor, which seems to have brought much of Shropshire and the Midlands
above sea-level, where the Cambrian and Pre-Cambrian rocks suffered
erosion. This episode may be regarded as the precursor of the great 'Caledonian' mountain-building movements which reached their maximum in
post-Silurian time.

At the beginning of Ordovician time progressive subsidence again began
in this district, and the basal Stiperstones Quartzite represents the sandy
deposit of a shallow sea which was subsequently followed by the grits, flags
and shales of the lower part of the Ordovician. At this time also there was a
great outburst of volcanic activity, both submarine and subaerial, producing
the tuffs and lavas of the Shelve district. Further subsidence carried the sea
into the eastern parts of Shropshire and the Upper Ordovician (Caradocian)
grits, sandstones and limestones were laid down unconformably on Cambrian
and Pre-Cambrian rocks, while deposition of shales and ashes proceeded
without break in the western areas.

Aerial view of the Church Stretton Valley and neighbouring hills looking southwards from near Condover (*Cambridge University Aerial Photograph No. Y 84*)     **Plate II**

*(For full explanation see p. ix)*

Extensive earth movements at the beginning of Silurian times resulted in a considerable tract of land appearing over Shropshire and the Midlands. This was subjected to denudation while Lower Llandovery deposits were forming on the sea floor to the west. When subsidence again brought the sea over this land Middle and Upper Llandovery sediments were laid down unconformably upon all the subdivisions of the Ordovician and on the Cambrian and Pre-Cambrian rocks of Shropshire and the Malvern district. Relics of coastal features formed during the initial Llandovery transgression are locally well preserved.

Throughout Upper Llandovery times the area of the sea extended and the depth of water increased, and, by middle Wenlock times, deposition was continuous over the Welsh Borderland in moderately deep water. The period of maximum transgression was followed in late Wenlock and throughout Ludlow times by a long period during which a 'basin' area of fairly deep water and rapid sea-floor subsidence to the north-west of a line from Horderley to Radnor merged into a 'shelf' area to the south-east where subsidence was slow and water much shallower. A reef facies of the Woolhope Limestone had formed against the Old Radnor ridge prior to the period of maximum transgression, and the shallowness of the 'shelf' sea during late Wenlock times is indicated by the formation of 'reef-knolls', masses of coral growing in warm and well-aerated water.

The close of Silurian times in north-west Britain was marked by the great earth movements, accompanied by volcanic episodes, of the 'Late Caledonian' mountain-building epoch. Within the Welsh Borderland, however, sedimentation went on continuously from the Silurian to the end of Lower Old Red Sandstone times. The conditions of sedimentation changed considerably at the end of the period of deposition of the Ludlow Beds. The shelly limestones and shales of definitely marine type gave place to the red marls and sandstones, with cornstone bands, of the Lower Old Red Sandstone, which appear to have been deposited under estuarine or deltaic conditions at no great distance from the land areas which emerged as the result of the earth movements mentioned above. From these rising continental masses vast quantities of mud and sand were swept down and spread out in the shallow waters, where continuous down-warping kept pace with sedimentation.

Upward movement accompanied by folding, and followed by much denudation, took place at the end of Lower Old Red Sandstone time in the Welsh Borderland, and no deposits of Middle Old Red Sandstone age appear to have been formed. This was followed by depression resulting in the deposition of the Upper Old Red Sandstone in unconformable relation to various horizons of the Lower Old Red Sandstone.

At the close of the Old Red Sandstone period a great part of the continental areas had been reduced to sea-level, and with further subsidence a fresh invasion of the sea took place over most of the British area.

In this sea the marine deposits of the Lower Carboniferous were formed, consisting in large part of limestones and shales. There were occasional episodes of volcanic activity at this period.

At the close of Lower Carboniferous time much terrigenous material was swept into areas adjacent to this region, converting them into delta swamps. Great areas of waterlogged silt were periodically colonized by exuberant plants of the horsetail, fern and club moss families. Peaty morasses formed in these areas covered with the dense vegetation which eventually decayed and became coal. Intermittent subsidence gave rise to thick deposits of sand and mud, forming the measures between the successive layers of coal, or occasionally to incursions of the sea resulting in 'Marine Bands' which are found at intervals in the Coal Measures.

Important earth movements, foreshadowing greater movements to come, took place in the district before the deposition of the Upper Coal Measures, and were accompanied by intrusions and extrusions of basic lava. The Upper Coal Measures rest unconformably on the pre-Carboniferous rocks in the Shrewsbury coalfields and on the folded older Coal Measures of the Coalbrookdale and Wyre Forest areas on the east of the region.

With the close of the Carboniferous period the main 'Hercynian' crustal movements set in bringing strong compressive forces from the south. Possibly because of the N.N.E.–S.S.W. trend of the margin of the resistant mass lying west of the Church Stretton Fault the sharpest folding in response to these pressures was along mainly north to south axes such as those of Malvern, Usk and Woolhope.

The mountainous terrain which resulted from the main Hercynian earth movements was denuded under continental conditions during Permian times by a climatic environment which favoured deep oxidation of much of the land surface. Red oxides of iron were synthesized in abundance and outcrops of coal seams were either burnt or converted to carbonates by complex physico-chemical processes. The deeply weathered terrain was then gradually buried during early Triassic times by mainly sandy sediment, some of which may have accumulated under arid or semi-arid conditions. The main depressions filled by the sandy Bunter sediments were later inundated by brackish water from an extensive land-locked sea, and, during later Keuper times, red muds with evaporite minerals accumulated in these areas. It is uncertain whether more than marginal areas of the Welsh Borderland Region were involved in the transgression of the Keuper sea.

The history of the rock formations of the region ends with the Trias, all later stratified deposits having been removed by denudation; and there only remain to be mentioned the deposits of the Glacial and post-Glacial periods.

Ice moving from south-west Scotland down the Irish Sea advanced over the North Shropshire plain probably during two periods within the last 100 000 years, and pressed against the northern side of the old Palaeozoic mass of south Shropshire, while Welsh ice descended from the hills down the Severn Valley and farther south into the valleys of the Teme and Wye. A discussion of these episodes is given in more detail later.

On the retreat of the ice tumultuous deposits of sand, gravel and boulder-clay were left behind, masking the solid rocks over large areas. The rivers were, in many cases, now obliged to find new channels, and the present system of drainage is to a large extent a legacy of the Pleistocene glaciations.

## Geological Sequence

An outline of the geological succession represented in the Welsh Borderland region above the Pre-Cambrian is given in Fig. 3. The Pre-Cambrian rocks are the subject of the next chapter.

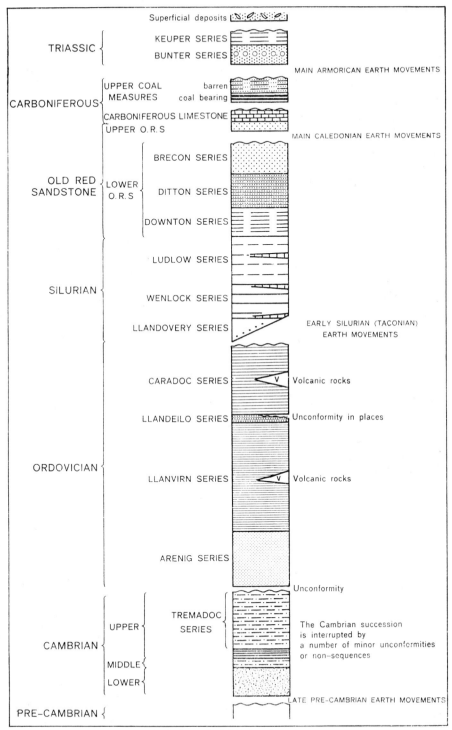

FIG. 3. *Generalized section of the Phanerozoic rock succession in the Welsh Borderland*

# 2.  Pre-Cambrian

Some of the oldest rocks in England and Wales appear at the surface in the Welsh Borderland (Figs. 2, 4). As most of them are hard and resistant to weathering, these ancient, Pre-Cambrian, rocks commonly form high ground including such well-known hills as the Wrekin, Caer Caradoc and the Long Mynd in Shropshire, and the Malvern Hills.

Since these rocks have been subjected to the vicissitudes of a long period of time, they are all more or less altered; some so much that their original character is almost obliterated. In most cases, however, they are not so changed that their mode of origin cannot readily be inferred.

No indisputable fossil remains or traces have yet been found in these Pre-Cambrian rocks. They cannot, therefore, be subdivided or correlated from place to place by means of fossils as can the Cambrian and later systems. Instead, a grouping based on lithological characters has to be adopted. Also, since the rocks have been considerably disturbed and their original relations to one another altered or rendered obscure, the relative age of the groups is in some cases in doubt. Isotopic age-determinations are, as yet, very few in number though it is possible that they may provide a means of correlation in the future. In many cases the Pre-Cambrian age of these rocks is proved by the fact that Lower Cambrian strata rest unconformably on them. Sometimes, however, such direct evidence is not available and the Pre-Cambrian age is inferred on other grounds.

The Pre-Cambrian rocks of the Welsh Borderland are grouped as follows, the various divisions not necessarily being in descending order of age:

Longmyndian
    Wentnor Series (Western Longmyndian) up to about 12 000 ft (3700 m) thick
    Stretton Series (Eastern Longmyndian) up to about 14 000 ft (4300 m) thick
Uriconian
    Eastern Uriconian lavas and tuffs
    Western Uriconian lavas and tuffs
?Uriconian
    Warren House Group of Malvern; lavas and tuffs
    Igneous rocks of Stanner Rocks, Worsel Wood and Hanter Hill
Rushton Schists
Primrose Hill Gneisses and Schists
Malvernian; gneisses and schists

All the principal classes of rocks are represented in these Pre-Cambrian groups. Some of them are sedimentary rocks, composed of the debris of pre-existing strata deposited in water as mud, sand or gravel, in more or less horizontal layers, and subsequently consolidated, deformed and dislocated. Such rocks are represented by the Longmyndian, named after the Long Mynd, near Church Stretton, and by similar strata in Radnorshire and Gloucestershire.

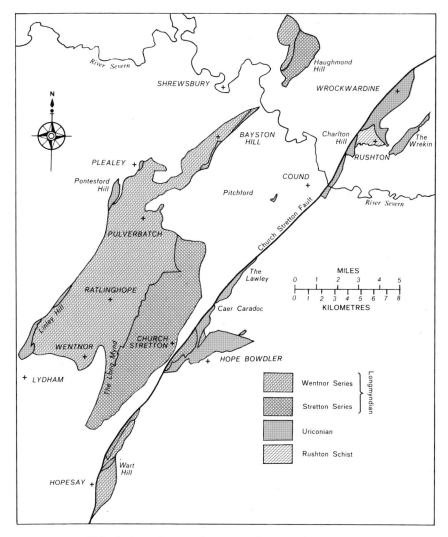

FIG. 4. *Distribution of Pre–Cambrian rocks in Shropshire*

Others of the Pre-Cambrian rocks are of volcanic origin. These include ancient lavas, and tuffs and volcanic breccias, that is, rocks consisting of material ejected as dust or fragments from a volcanic vent and deposited on land or in water, either directly or after having been washed down the slopes of a volcano. Such rocks are represented by the Uriconian of Shropshire; named by Callaway after the Roman city of Uriconium, the name of which was probably derived from that of the Wrekin which is mainly formed of these volcanic rocks. Caer Caradoc and other hills near Church Stretton and Pontesford Hill near Pontesbury, are also formed of Uriconian rocks, and there are similar rocks in the Malvern Hills and near Old Radnor.

Rocks that since their formation have been subjected to elevated temperature and pressure whereby minerals and textures not originally present have been formed, are known as metamorphic rocks. Before metamorphism many of these were of intrusive igneous origin, that is they were formed as injections of molten matter into pre-existing rocks amongst which they cooled and crystallized. Others were of sedimentary origin. Metamorphic rocks are represented by the Malvernian of the Malvern Hills and also by the Rushton Schists and Primrose Hill Gneisses and Schists of Shropshire.

In addition to those already mentioned, igneous rocks occur amongst the other types, into which they were intruded in a molten state. These include pink or light coloured rocks rich in silica (acid rocks) and dark coloured rocks poor in silica (basic rocks). The acid intrusions are probably of nearly the same age as the volcanic rocks with which they are associated, but some of the basic intrusions may be of considerably later date.

The relative age of the various Pre-Cambrian groups of the Welsh Borderland has been a source of controversy for many years and it still presents a number of unsolved problems. The metamorphic rocks (Malvernian, Rushton Schists, Primrose Hill Gneisses and Schists) are almost certainly the oldest rocks in the region, but their relation to each other still remains unknown. It is now generally agreed that the Eastern and Western Uriconian are approximately contemporaneous. From included fragments it is evident that they were partly derived from, and are therefore younger than, a metamorphic complex which included rock types similar to those in the Malvernian, Rushton Schists and the Mona Complex of Anglesey. The Longmyndian sediments contain abundant derived fragments of both metamorphic and Uriconian rock types. It has recently been shown (p. 19), that the Wentnor Series rests unconformably on the Western Uriconian, while the Stretton Series is undoubtedly younger than the Eastern Uriconian (p. 22). Consequently, it appears that the whole of the Longmyndian is younger than the Uriconian.

Although it is not possible to reconstruct the palaeogeography of Pre-Cambrian times, an attempt can be made to piece together a sequence of events during the later Pre-Cambrian. The Uriconian volcanic suite is typical of post-orogenic vulcanism, while the Longmyndian sediments, which appear to have been laid down in a fault-bounded trough, have the characteristics of post-orogenic molasse deposits (continental clastic sediments). It is therefore not unreasonable to suggest that the Uriconian and Longmyndian may represent post-orogenic phases of a Pre-Cambrian orogenic cycle which gave rise to an orogenic belt of which the metamorphic rocks of the Malvernian and the Mona Complex may have formed a part.

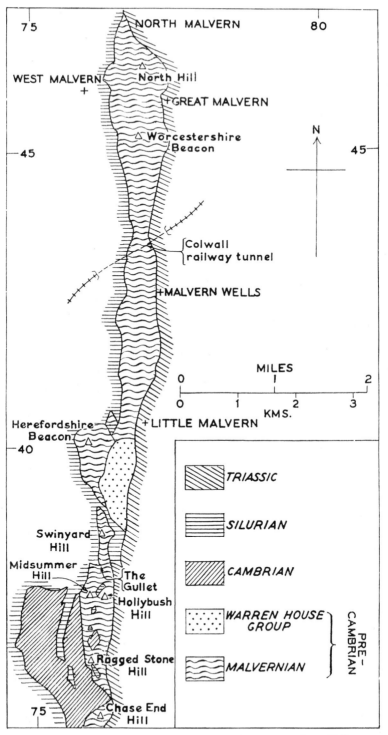

FIG. 5. *Distribution of Pre-Cambrian rocks in the Malvern Hills*
(After Groom 1910.)

## Malvernian

The Malvernian gneisses and schists form the central axis of the Malvern Hills, a prominent north to south ridge some 8 miles (13 km) long (Fig. 5). They include a great variety of igneous and metamorphic rocks, poorly exposed over much of their outcrop, which are generally considered to be some of the oldest rocks in the Welsh Borderland.

The extremely altered and mylonitized state of much of the Malvernian, coupled with its poor exposure, has made it a difficult group to study. Widely differing interpretations of its original characteristics and subsequent metamorphic and tectonic history have been put forward. No detailed petrological work has been published since that of Callaway (1893). Brammall, whose work remains unpublished, considered that the Malvernian gneisses and schists resulted from the metamorphism of a series of sedimentary and interbedded igneous rocks to schists, epidiorites and amphibolites followed by further, high-grade, metamorphism to rocks such as quartz-mica-schists, hornblende-biotite-gneiss and metasomatic granite and syenite. This metamorphism and migmatization, most strongly developed at the northern end of the range, was followed by the injection of pink granite and pegmatite, extensive faulting and thrusting and the intrusion of dolerite dykes. On the other hand recent work by Lambert and Rex (1966) suggests that true gneisses and schists of high grade are not present and that the Malvernian 'gneisses' have resulted from hydrothermal alteration and shearing of igneous rocks such as quartz-diorite and amphibolite. They consider that schists are rocks largely confined to shear zones and fault blocks.

The rocks at the northern end of the range, on North Hill and Worcestershire Beacon, are largely of igneous aspect and include relatively unfoliated diorite, quartz-diorite and hornblende-granite. Part of the Colwall railway tunnel (Fig. 6) was driven through altered diorite (epidiorite) though the rocks at the surface above the tunnel are chiefly quartz-mica-schist. Farther south, on Herefordshire Beacon, the rocks are mainly hornblende- and mica-gneisses, though at the northern end of the hill a basalt has been recognized. Swinyard Hill also shows mica-gneiss, with diorite, more or less altered, and hornblende-granite. On Midsummer Hill and Hollybush Hill both gneisses and schists occur, whilst schists prevail on Ragged Stone Hill where they include a brecciated quartzite. Chase End Hill, at the southern end of the range, is formed mainly of hornblende-gneiss.

Pegmatite veins are common and are particularly well developed in the Gullet Quarry, near Hollybush Hill. Good examples of basic intrusions (mainly dolerite) may be seen at the Dingle Quarry, West Malvern, at the Tank Quarry, North Hill and at the Gullet Quarry.

A very small area of supposed Malvernian rocks occurs about $\frac{1}{2}$ mile (0·8 km) west-south-west of Martley, Worcestershire, on the line of the Malvern–Abberley Axis. The rocks comprise gabbro and crushed granite and are associated with a quartzite which may be of Cambrian age (Mitchell and others 1962).

## Primrose Hill Gneisses and Schists

The Primrose Hill Gneisses and Schists crop out over an area about 350

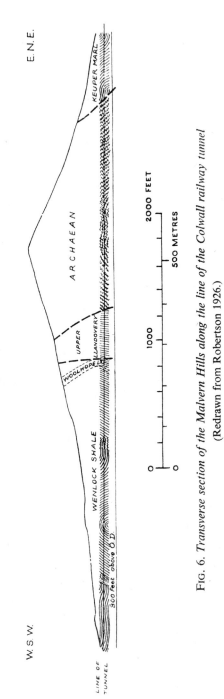

FIG. 6. *Transverse section of the Malvern Hills along the line of the Colwall railway tunnel* (Redrawn from Robertson 1926.)

yd (320 m) wide on Little (Primrose) Hill at the southern end of the Wrekin (Fig. 7). They are overlain unconformably by the Wrekin Quartzite and are faulted against the Uriconian of the Wrekin by the Primrose Hill Fault. It is possible that a continuation of this fault to the north-west also forms the north-eastern limit of the Rushton Schists. The rocks can be divided into an upper group composed of acid igneous material with a strong cataclastic structure and lower groups of gneissose and schistose types. All three groups are veined with pink granophyric material. Pocock and others (1938) concluded that all the Primrose Hill rocks could have been formed from the Uriconian volcanics by injection of granophyre coupled with dynamic metamorphism. However, it now seems probable that at least the lower two groups, which include such rocks as hornblende-schists and biotite-gneiss, are pre-Uriconian and have close affinities with the Malvernian.

## Rushton Schists

These rocks, first described by Callaway (1884), have a limited outcrop between Rushton and Uppington (Fig. 4). They are overlain unconformably by the Cambrian Wrekin Quartzite, but their relation to the adjacent Uriconian rocks is everywhere concealed. The rocks are quartz-mica-schists with garnet and, generally, much epidote. They are cut by a basic dyke, itself schistose, and by later felsite dykes similar to those in the Uriconian. It has been suggested that the Rushton Schists were formed by the local metamorphism of Uriconian or Longmyndian rocks, but recent work (Dearnley *in* Greig and others 1968, pp. 72–3) shows that they are low grade regional metamorphic rocks (greenschist facies) which closely resemble some parts of the Mona Complex of Anglesey.

## Uriconian

These rocks have been separated into two groups, the Eastern and Western Uriconian, which crop out respectively to east and west of the Long Mynd and its northward prolongations (Fig. 4). The Eastern Uriconian outcrops lie along the line of the Church Stretton Fault Complex and the Western Uriconian along the sub-parallel Pontesford–Linley fault zone.

Petrographically the Eastern and Western Uriconian are very similar. They consist of a calc-alkaline suite of lavas of basic (basalt), intermediate (andesite, dacite) and acid (rhyolite) types with associated tuffs, and intrusions of basic (dolerite) and acid (granophyre, rhyolite) rocks. Many of the rhyolitic lavas show flow-banding and some have a spherulitic structure resulting from recrystallization from an original glassy condition.

The Uriconian volcanic suite is typical of the last phase of igneous activity of an orogenic cycle. Derived fragments in the Uriconian tuffs show that they are later than a series of low grade metamorphic rocks (greenschist facies) similar to the Rushton Schists and some of the rocks of the Mona Complex of Anglesey (p. 11).

### Eastern Uriconian

The Eastern Uriconian rocks are developed in two main areas, the sub-parallel Wrekin and Wrockwardine outcrops at the northern end of the

Church Stretton Fault Complex and the Church Stretton area farther to the south.

The Wrekin range, type area of the Uriconian, comprises the Ercall at the northern end, Lawrence Hill, the Wrekin, and Primrose (Little) Hill at the south (Fig. 7). The Uriconian rocks have a general north-eastward dip and are overlain unconformably by the basal Cambrian Wrekin Quartzite.

At the north end of the Ercall there is an assemblage of tuffs, mainly rhyolitic. The central part of the hill is formed by a granophyre (a rock consisting mainly of a micrographic intergrowth of quartz and feldspar) which may be intruded into the volcanic rocks. South of the granophyre more tuffs crop out, followed by a group of rhyolite lavas in the glen between the Ercall and Lawrence Hill. Nodular and flow-banded rhyolite can be seen in the crag on the north side of the glen. The quarry in Forest Glen, between Lawrence Hill and the Wrekin, shows a fine section of tuffs and lavas including a bed of tuff with rounded boulders of banded rhyolite. The highest part of the Wrekin is occupied by another group of rhyolite lavas, of which banded examples can be seen in the crags near the summit. On the south-west slopes tuffs, breccias and lavas of somewhat basic composition crop out in two bands, separated by a band of predominantly rhyolite rocks. Intrusions of dolerite and quartz-dolerite are well displayed in the Forest Glen and on the northern slopes of the Wrekin. These basic dykes cut across the bedding of the volcanic rocks but do not penetrate the overlying Cambrian.

Typical Uriconian rocks are also present in the Wrockwardine and Charlton Hill areas (Fig. 4). Around Wrockwardine there are green basic tuffs and a group of red and purple rhyolites with flow-banding and spherulitic structure. These rhyolites are well seen at Overley Hill (Lea Rock). At Charlton Hill there are tuffs and andesite and also a conglomerate with well-rounded pebbles which include rock types similar to the Primrose Hill Gneisses (p. 15) and the Ercall granophyre. From this latter occurrence Callaway (1891) concluded that the Uriconian rocks were younger than the granophyre. However, as already mentioned, the Ercall granophyre may be intrusive into the Uriconian rocks of that hill and therefore later in date. If this latter interpretation is correct the pebbles in the Charlton Hill conglomerate might have been derived from an earlier granophyric mass or, possibly, the Charlton Hill Uriconian rocks were formed somewhat later than those of the Ercall.

Uriconian rocks form the high ground, including such well-known hills as Caer Caradoc and the Lawley, on the eastern side of the Church Stretton Valley between Leebotwood and Little Stretton. The rocks are mainly lavas and tuffs, with associated intrusions, similar to those of the type area of the Wrekin. They occur in five distinct areas (the Lawley; Caer Caradoc Hill, Helmeth Hill and Ragleth Hill; Hazler Hill; Hope Bowdler Hill; Cardington Hill) separated by faults. Although a stratigraphical sequence of rocks has been worked out in some of these areas it has not proved possible to correlate these sequences from one area to another (Greig and others 1968). Basic tuffs and amygdaloidal rhyolites and andesites are present on the Lawley, with basic intrusions including olivine-dolerite. The tuffs and lavas of the Caer Caradoc area (Fig. 8), which are more than 4000 ft (1200 m) thick, are

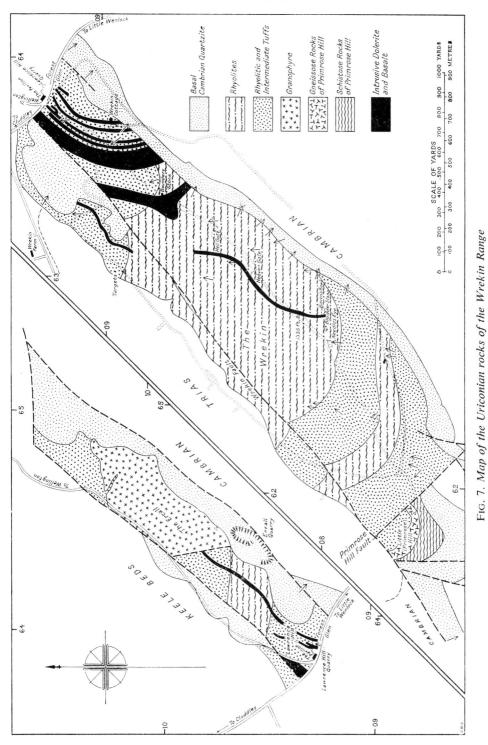

Fig. 7. *Map of the Uriconian rocks of the Wrekin Range*
(From Pocock and others 1938.)

particularly well exposed near the summit of the hill. They are apparently folded on north-westerly axes (see p. 89) and appear to be overlain unconformably by the Longmyndian Helmeth Grit. The Hazler Hill area is largely occupied by basaltic lavas with interbedded tuffs. These beds are overlain unconformably to the south by the Ordovician Harnage Shales, which also form the Neptunian dykes within the Uriconian of Hazler Quarry (Strachan and others 1948). Andesites, conglomerates and rhyolites are present on Hope Bowdler Hill, and the Cardington Hill area displays a sequence, about 4000 ft (1200 m) thick, of tuffs, andesites and dacites with associated intrusions of quartz-porphyry, quartz-microdiorite and dolerite.

A small faulted inlier of Uriconian rocks, within the Church Stretton Fault Complex, forms the prominent knoll of Wart Hill about 1 mile (1·6 km) north-north-east of Hopesay (Fig. 4).

### Western Uriconian

Western Uriconian rocks crop out at intervals between Plealey and Lydham (Fig. 4) along the line of the Pontesford–Linley fault zone.

At Plealey, lenticular outcrops of felsite are found adjacent to rocks of Longmyndian aspect. Farther south, at the Lyd Hole in the Habberley Brook, near Pontesford, an alternation of basic and acid tuffs and lavas dips to the south-east at a high angle. They appear to be faulted against Longmyndian grit to the east (Dean and Dineley 1961).

Pontesford and Earl's hills (Plate VB and Fig. 9) are formed of a succession of south-easterly dipping volcanic rocks intruded by large masses of olivine-dolerite. The volcanic rocks begin, at the north end, with a group of rhyolites which are locally flow-banded or nodular. These are succeeded by rather acid tuffs and a small outcrop of basalt lava. Much of Pontesford Hill is occupied by a dolerite intrusion, with a group of basic tuffs to the south and the west. These include coarse breccias with angular lava fragments, and tuffs consisting largely of altered glass of chemical composition similar to that of basalt (palagonite tuffs). The highest part of Earl's Hill is formed by intrusive dolerite; on the western side of the hill there are basic tuffs, some with palagonite, others very fine-grained and flinty in appearance (halleflintas), with flows of basaltic lava. Similar rocks crop out in a small ridge north-east of Habberley. On the south-east side of Earl's Hill another group of acid rocks occurs which have been described as rhyolitic breccias and tuffs, and rhyolites. The latter, which locally have a spherulitic appearance, have recently (Dearnley 1966) been shown to be devitrified welded tuffs.

The most southerly outcrop of the Western Uriconian is almost continuous from a point about 2 miles (3·2 km) west-north-west of Ratlinghope southwards to about 1 mile (1·6 km) north of Lydham. In the northern part of this outcrop, around The Knolls, the rocks are rhyolites, followed eastward by an alternation of coarse and fine-grained tuffs of intermediate or basic composition. They dip westwards and are in contact with Longmyndian grit and conglomerate to the east. The nature of the contact here is uncertain. To the west they are faulted against Upper Cambrian (Tremadoc) shales.

Farther south, in Chittol Wood, 1½ miles (2·4 km) north-west of Norbury, the volcanic rocks are intruded by basalt and granophyre. Similar rocks, together with tuffaceous shales and many basic intrusions, crop out in

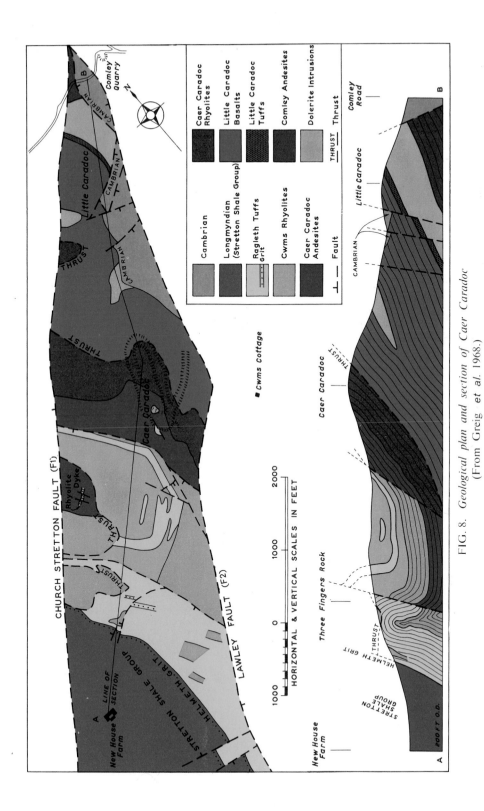

FIG. 8. *Geological plan and section of Caer Caradoc*
(From Greig *et al.* 1968.)

Linley Big Wood, north of Lydham. The nature of the contact between the Uriconian and Longmyndian rocks of the Chittol is of great significance. James (1952, 1956) has shown that Wentnor Series (Western Longmyndian) rocks rest unconformably on the Uriconian and that this unconformity has later been inverted so that the Uriconian now appears to overlie the Longmyndian (Fig. 10). Consequently the Western Uriconian must be older than the Wentnor Series (see p. 21).

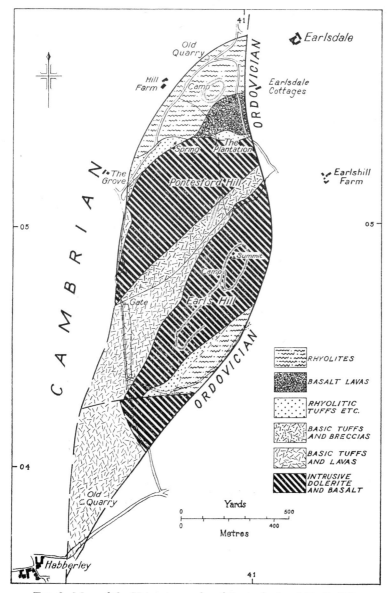

FIG. 9. *Map of the Uriconian rocks of Pontesford and Earl's hills*
(From Pocock and others 1938.)

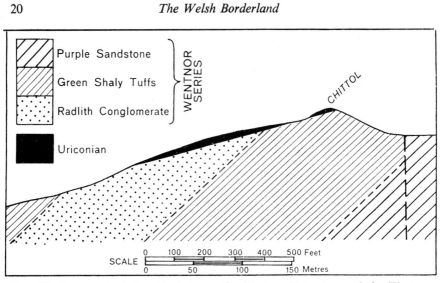

FIG. 10. *Structural relationship between the Western Uriconian and the Wentnor Series (Western Longmyndian) at Chittol, Shropshire*

The Wentnor Series was laid down unconformably on the Uriconian, the whole sequence subsequently being inverted.

(From James 1956.)

## ?Uriconian

### Warren House Group of Malvern

This group of volcanic rocks bears a strong resemblance to the Uriconian of Shropshire with which it may probably be correlated. It crops out on Hangman's Hill, Broad Down and Tinker's Hill, to the east and south-east of Herefordshire Beacon (Fig. 11). It includes lavas rich in silica (sodic and epidotized rhyolites), others poor in silica (spilites), and tuffs, together with dolerites which are probably intrusive. The dip seems to be eastward and these rocks appear to lie unconformably on the Malvernian. The Warren House Group is considered to be Pre-Cambrian partly on account of its resemblance to the Uriconian of Shropshire and partly because rolled pebbles of very similar rocks occur in the Malvern Quartzite, at the base of the Cambrian.

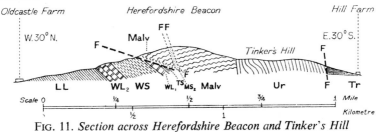

FIG. 11. *Section across Herefordshire Beacon and Tinker's Hill*

Tr = Trias; LL = Lower Ludlow Shales; $WL_2$ = Wenlock Limestone; WS = Wenlock Shale; $WL_1$ = Woolhope Limestone; TS = Woolhope ('Tarannon') Shale; $MS_2$ = Upper part of May Hill Sandstone; Ur = Uriconian; Malv = Malvernian; F,F = Faults.

(From Groom 1900.)

**Igneous rocks of Hanter and Stanner**

The hills of Stanner Rocks, Worsel Wood and Hanter Hill (Plate IIIA), near Old Radnor, lie within the southward continuation of the Church Stretton Fault Complex. They are composed of basic igneous rocks (dolerite and gabbro) with dykes of various acid rocks including quartz-porphyry, felsite and granophyre. Some of the latter closely resemble the granophyres associated with the Uriconian of the Wrekin and north of Lydham, and the Hanter and Stanner igneous rocks were, in part, tentatively referred to the Uriconian by Callaway (1900). Some later authors considered that they were intrusive into the surrounding Silurian shales, but more recent work (Holgate and Knight-Hallowes 1941) indicates that most of the boundaries are faulted though it is possible that locally there is an unconformity between the igneous rocks and the Silurian. These more recent studies have emphasized the resemblance of the igneous rocks to some of the Pre-Cambrian intrusions in Shropshire and have shown that very similar rock types occur as pebbles in the Longmyndian conglomerates of the nearby Old Radnor area. There is thus a strong presumption that the Hanter and Stanner rocks are Pre-Cambrian, probably Uriconian, in age.

# Longmyndian

The Longmyndian is most fully developed on the Long Mynd itself and in the area to the west around Wentnor and Bridges. This outcrop extends northwards to Pulverbatch and Bayston Hill, with an isolated area north of the River Severn at Haughmond Hill (Fig. 4). In Shropshire, other outcrops of Longmyndian rocks occur at Pitchford and within the Church Stretton Fault Complex between Caer Caradoc and Hopesay hills. Farther south, rocks of Longmyndian type occur at Pedwardine, Herefordshire, Old Radnor and possibly at Huntley Quarry, Gloucestershire.

The Longmyndian has been divided into two series, the Wentnor Series (Western Longmyndian) and the Stretton Series (Eastern Longmyndian). On and to the west of the Long Mynd, the whole sequence has a general steep westerly dip, but further evidence (James 1956) indicates that the Wentnor Series is folded into a major isoclinal syncline with an inverted western limb (Fig. 45). On this western limb the Wentnor Series rests unconformably on the Western Uriconian (Fig. 10 and p. 19) and it appears that the Stretton Series has been cut out by an unconformity at the base of the Wentnor Series.

The sequence is almost entirely sedimentary, with a range in rock types from banded argillaceous rocks to coarse sandstones and conglomerates. Subgreywacke-sandstones occur commonly, especially in the Wentnor Series, and there are beds of tuff in the Stretton Series. Many types of sedimentary structures occur in the Longmyndian including current and graded bedding, ripple marks, mud cracks and rain prints. The supposed Longmyndian fossils, including '*Arenicolites*', are now thought to be of inorganic origin (Greig and others 1968). The rock types and sedimentary structures in the Longmyndian indicate deposition in shallow water, and deposition apparently took place in an elongate crustal depression possibly related to the Pontesford–Linley and Church Stretton fault systems.

**Stretton Series (Eastern Longmyndian)**

This Series, approximately 14 000 ft (4300 m) thick, consists mainly of mudstones and siltstones with subordinate, sometimes massive, sandstones and some beds of tuff. The rocks, with a general steep westerly dip, are well displayed on the eastern side of the Long Mynd in Ashes Hollow, the Cardingmill Valley and Lightspout Hollow. On the eastern side of the Church Stretton Valley the basal member of the Series, the Helmeth Grit, appears to rest unconformably on the Eastern Uriconian (Greig and others 1968, p. 76) although Cobbold and Whittard (1935) considered that the junction was transitional.

The Stretton Series has been divided into the following five groups in ascending order. The thicknesses of the various divisions are approximate.

1. *Stretton Shale Group.* This Group, perhaps 3000 ft (900 m) thick, comprises greenish grey mudstones and siltstones, often well laminated. They were subdivided by Lapworth (1910) into the Watling Shales on the east side of the Church Stretton Valley and the Brockhurst Shales, well exposed at the entrance to the Cardingmill Valley, on the west side; these terms have now fallen into disuse. The two outcrops are separated by the Church Stretton Fault and a narrow area of Silurian rocks. In the eastern outcrop four beds of tuffaceous grit (the Helmeth Grit) are developed within the basal 100 ft (30 m) of the sequence. They contain abundant fragments of rocks of Uriconian type and rest on the Eastern Uriconian of the Caer Caradoc area (see above). Several faulted strips of the Stretton Shale Group occur within the Church Stretton Fault Complex north of Wart Hill. The small area of shales and flags at Pitchford may perhaps be included in either the Stretton Shale Group or the Burway Group.

2. *Burway Group.* This Group is about 2000 ft (600 m) thick and consists of flaggy greenish grey laminated siltstones with some sandstone beds. At base there is a bed of silicified tuff (the Buxton Rock) up to 24 ft (7·3 m) thick, and at the top a series of massive sandstones 100 to 190 ft (30–58 m) thick (the Cardingmill Grit). The group extends along the eastern side of the Long Mynd and is well exposed in the sides of the main Long Mynd valleys such as Callow Hollow and Ashes Hollow; the Buxton Rock is best seen in Buxton Quarry, All Stretton. North of the Long Mynd, green shales and sandstones on the eastern side of Haughmond Hill probably belong to the Burway Group.

3. *Synalds Group.* The Synalds Group comprises 1600 to 2800 ft (490–850 m) of predominantly purple shaly mudstones and siltstones with subordinate sandstone beds. There are several thin beds of tuff; the most persistent beds, near the top of the Group, are known as the Batch Volcanics. The Synalds Group extends along the whole length of the Long Mynd and, like the Burway Group, is well exposed in the sides of the cross valleys.

4. *Lightspout Group.* This Group is about 1700 to 2700 ft (520–820 m) thick and consists mainly of flaggy greenish grey siltstones and sandstones with some massive sandstone beds especially in the lower part. These include the sandstones which form the Lightspout waterfall. The topmost beds of

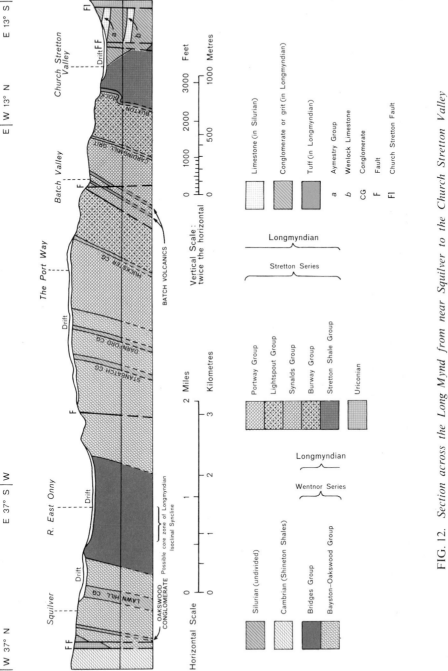

FIG. 12. *Section across the Long Mynd from near Squilver to the Church Stretton Valley (Mainly from One–Inch Geological sheet 166 (Church Stretton); western end based on James 1956)*

the Group are mainly purple in colour with at least one tuff bed, and they are very similar to strata in the Synalds Group.

5. *Portway Group.* Some 600 to 3500 ft (180–1070 m) thick, this Group comprises purple and greenish grey mudstones and siltstones with beds of sandstones and with the Huckster Conglomerate, up to 60 ft (18 m) thick, at the base. Its outcrop lies on the summit plateau of the Long Mynd, and exposure is poor. North of the Long Mynd purple and green sandy shales, probably belonging to the Portway Group, crop out in Smethcott Dingle; similar beds are seen on the east side of Sharpstone Hill and northwards to the River Severn. James (1956) considered that there was an unconformity between the Portway and Lightspout groups, and he classified the Portway Group as a separate series (the Mintonian). Later work (Greig and others 1968) has disproved the existence of such an unconformity and the Portway Group is regarded, as previously, as the topmost unit of the Stretton Series.

### Wentnor Series (Western Longmyndian)

This Series, approximately 12 000 ft (3700 m) thick, consists of purple sandstones, siltstones and mudstones with some thick beds of conglomerate. The Wentnor Series appears to rest unconformably on the Stretton Series. The main evidence for this unconformity is the greater geographical extent of rocks of Wentnor Series type and the absence of the Stretton Series to the west of the Long Mynd where Wentnor Series beds rest directly on the Western Uriconian. Also, at Haughmond Hill, the basal Wentnor Series Haughmond Conglomerate appears to rest unconformably on the lower part of the Synalds Group.

The Wentnor Series was divided by Whitehead (*in* Pocock and others 1938) into three groups; from east to west they were the Bayston, Bridges and Oakswood groups. However, it is now thought that the Bridges Group forms the core of an isoclinal syncline (Fig. 45) and that the Bayston and Oakswood groups are equivalent. The sequence is as follows:

1. *Bayston–Oakswood Group.* This Group, about 400 to 800 ft (120–240 m) thick, consists of coarse-grained purple sandstones with subordinate mudstones and siltstones and some thick conglomerate beds. The conglomerates contain many pebbles of Uriconian rocks and some of metamorphic rock types. There are two areas of outcrop corresponding to the old Bayston and Oakswood groups, separated by the outcrop of the Bridges Group.

The Bayston outcrop extends northwards from Asterton to Bayston Hill and Haughmond Hill. Three thick conglomerates, from east to west the Haughmond, Darnford and Stanbatch conglomerates, occur in this area. The Haughmond Conglomerate, unlike the other two, does not extend as far south as the Long Mynd.

The Oakswood outcrop extends from Linley to near Plealey. Three conglomerates, from east to west the Lawn Hill, Oakswood and Radlith conglomerates are also developed in this area but they probably do not correlate with those in the Bayston area to the east. The Lawn Hill Conglomerate is the thickest, reaching about 1300 ft (400 m) locally.

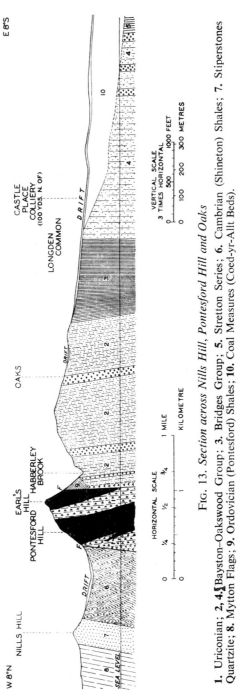

FIG. 13. *Section across Nills Hill, Pontesford Hill and Oaks*

**1.** Uriconian; **2, 4.** Bayston–Oakswood Group; **3.** Bridges Group; **5.** Stretton Series; **6.** Cambrian (Shineton) Shales; **7.** Stiperstones Quartzite; **8.** Mytton Flags; **9.** Ordovician (Pontesford) Shales; **10.** Coal Measures (Coed-yr-Allt Beds). Intrusive dolerite: black.    F, faults.

(From One-inch Geological Sheet 152 (Shrewsbury).)

2. *Bridges Group.* The Bridges Group, some 2000 to 4000 ft (600–1200 m) thick, comprises purple mudstones, siltstones and flaggy sandstones with some more massive sandstones. Sedimentary structures such as graded and current bedding within the laminated beds of this Group have provided much of the evidence for the isoclinal syncline in the Wentnor Series.

Purple grits and conglomerates of Wentnor Series type occur in a number of faulted inliers within the Church Stretton Fault Complex between Cwm Head and Hopesay Hill. Similar grits and conglomerates occur in inliers at Pedwardine, Herefordshire, and near Old Radnor. The inlier at Huntley Quarry, near May Hill, in Gloucestershire, shows grits and shales that are probably Longmyndian, though it is difficult to assign them to any of the groups recognized in Shropshire.

Basic intrusions in the Longmyndian rocks are mainly of quartz-dolerite type, though olivine-dolerite or basalt occurs on Haughmond Hill and at Plealey. The quartz-dolerites sometimes contain free quartz but are more often characterized by the presence of a microscopic intergrowth of quartz and feldspar (acid mesostasis) in the interstices between the large crystals. Two small bosses of microdiorite occur in the Synalds Group near Jinlye, All Stretton. The characteristics of the dolerites are identical with those of dykes and sills intruded into the Cambrian and Ordovician strata near Pontesbury, and they are probably post-Cambrian and pre-Upper Llandoverian in age.

# References

BLAKE, J. F. 1890. On the Monian and Basal Cambrian Rocks of Shropshire. *Quart. J. Geol. Soc. Lond.*, **46**, 386–420.

BOULTON, W. S. 1904. The Igneous Rocks of Pontesford Hill (Shropshire). *Quart. J. Geol. Soc. Lond.*, **60**, 450–86.

CALLAWAY, C. 1879. The Precambrian Rocks of Shropshire—Pt. I. *Quart. J. Geol. Soc. Lond.*, **35**, 643–62.

—— 1884. On a new Metamorphic Area in Shropshire. *Geol. Mag.*, (3), **1**, 362–6.

—— 1886. On some Derived Fragments in the Longmynd and Newer Archaean Rocks of Shropshire. *Quart. J. Geol. Soc. Lond.*, **42**, 481–5.

—— 1891. On the Unconformities between the Rock-systems underlying the Cambrian Quartzite in Shropshire. *Quart. J. Geol. Soc. Lond.*, **47**, 109–25.

—— 1893. The Origin of the Crystalline Schists of the Malvern Hills. *Quart. J. Geol. Soc. Lond.*, **49**, 398–425.

—— 1900. On Longmyndian Inliers at Old Radnor and Huntley, Gloucestershire. *Quart. J. Geol. Soc. Lond.*, **56**, 511–20.

COBBOLD, E. S. and WHITTARD, W. F. 1935. The Helmeth Grits of the Caradoc Range, Church Stretton; their Bearing on Part of the Pre-Cambrian Succession of Shropshire. *Proc. Geol. Assoc.*, **46**, 348–59.

DEAN, W. T. and DINELEY, D. L. 1961. The Ordovician and Associated Pre-Cambrian Rocks of the Pontesford District, Shropshire. *Geol. Mag.*, **98**, 367–76.

DEARNLEY, R. 1966. Ignimbrites from the Uriconian and Arvonian. *Bull. Geol. Surv. Gt Brit.*, No. 24, 1–6.

GREIG, D. C., WRIGHT, J. E., HAINS, B. A. and MITCHELL, G. H. 1968. Geology of the country around Church Stretton, Craven Arms, Wenlock Edge and Brown Clee. *Mem. Geol. Surv.*

GROOM, T. T. 1900. On the Geological Structure of Portions of the Malvern and Abberley Hills. *Quart. J. Geol. Soc. Lond.*, **56**, 138–97.

—— 1910. The Malvern and Abberley Hills and the Ledbury District. Geology in the Field (Jubilee Volume of the Geologists' Association), 698–738.

HOLGATE, N. and KNIGHT-HALLOWES, K. A. 1941. The Igneous Rocks of the Stanner–Hanter District, Radnorshire. *Geol. Mag.*, **78**, 241–67.

JAMES, J. H. 1952. Notes on the relationship of the Uriconian and Longmyndian Rocks near Linley, Shropshire. *Proc. Geol. Assoc.*, **63**, 198–200.

—— 1956. The structure and stratigraphy of part of the Pre-Cambrian outcrop between Church Stretton and Linley, Shropshire. *Quart. J. Geol. Soc. Lond.*, **112**, 315–37.

LAMBERT, R. ST. J. and REX, D. C. 1966. Isotopic Ages of Minerals from the Pre-Cambrian Complex of the Malverns. *Nature, Lond.*, **209**, 605–6.

LAPWORTH, C. and WATTS, W. W. 1910. Geology in the Field, Shropshire. *Geol. Assoc. Jubilee Vol.*, 739–69.

MITCHELL, G. H., POCOCK, R. W. and TAYLOR, J. H. 1962. Geology of the country around Droitwich, Abberley and Kidderminster. *Mem. Geol. Surv.*

POCOCK, R. W., WHITEHEAD, T. H., WEDD, C. B. and ROBERTSON, T. 1938. Shrewsbury District including the Hanwood Coalfield. *Mem. Geol. Surv.*

ROBERTSON, T. 1926. The Section of the New Railway Tunnel through the Malvern Hills at Colwall. *Sum. Prog. Geol. Surv.* for 1925, 162–73.

STRACHAN, I., TEMPLE, J. and WILLIAMS, A. 1948. The Age of the Neptunian Dykes at Hazler Hill (Shropshire). *Geol. Mag.*, **85**, 276–8.

WHITEHEAD, T. H. 1955. The Western Longmyndian Rocks of the Shrewsbury District. *Geol. Mag.*, **92**, 465–70.

# 3. Cambrian

The Cambrian rocks of the Welsh Borderland occur in two principal areas. One is associated with the Eastern Uriconian Axis of Shropshire and extends from Lilleshall and the Wrekin (Fig. 15) (Cobbold and Pocock 1934; Pocock and others 1938) to the Caradoc, Comley and Cardington district (Fig. 16) (Cobbold 1921, 1927; Greig and others 1968), and the other with the ancient Malvernian axis of Herefordshire and Worcestershire (Fig. 19) (Groom 1899, 1902). Smaller areas of these rocks are also found between Pontesbury and Lydham on the western side of the Longmyndian Ridge of Shropshire and at Pedwardine, Herefordshire (Cox 1912), close to the line of the Church Stretton Fault.

These rocks are of great interest as being the most ancient sediments in this country to contain a recognizable sequence of fossils (Fig. 14); nevertheless, the variety of animal life which is represented in this oldest Cambrian fauna makes it certain that the ancestors of these forms must have existed in Pre-Cambrian times. The types of trilobites, brachiopods and pteropods which are the main constituents of the fauna in the oldest Cambrian rocks show a remarkable similarity to one another in widely separated areas, a fact which suggests that the extension of the sea at this period was world wide, and that similarity of conditions allowed the dispersal of the same types over a great part of the marine areas of the world.

The Cambrian sediments of the Welsh Borderland can be classified into four main divisions with distinct faunal characteristics:

*Lower Cambrian*, characterized by Olenellid trilobites (e.g. *Callavia*, Fig. 14A) and comprising a quartzite at the base overlain by glauconitic sandstones containing beds of shale and locally with a thin development of fossiliferous sandy limestones at the top.

*Middle Cambrian*, with the trilobites *Paradoxides* (Fig. 14B) and *Kootenia* [*Dorypyge*], consisting of shales and glauconitic sandstones with local thin conglomerates and lenticular limestones.

*Upper Cambrian* (excluding the Tremadoc Series), with Olenid trilobites and the brachiopod *Orusia lenticularis*, consisting mainly of shales with calcareous concretions.

*Upper Cambrian* (Tremadoc Series), with Olenid and other trilobites (e.g. *Shumardia pusilla*, Fig. 14H) some showing Ordovician affinities, as well as dendroid graptolites such as *Dictyonema flabelliforme* (Fig. 14E). It comprises grey shales with sandstone beds in the upper part.

A detailed zonal subdivision of the system has been made, based mainly on trilobites. The scattered nature of the Cambrian outcrops prevents any detailed delimitation of the palaeogeography of the period. However, it appears that throughout Cambrian times the Welsh Borderland was in a shelf area to the east of a subsiding geosyncline in Wales. The sea was never very deep and there were many breaks in deposition. The period began with

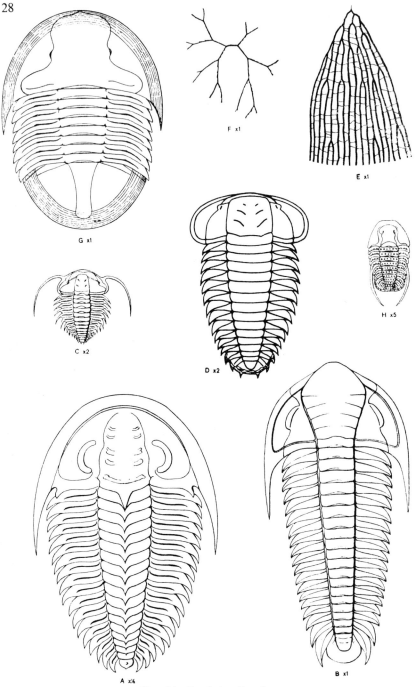

FIG. 14. *Cambrian Fossils*

**A.** *Callavia callavei* (Lapworth) × ½, Lower Cambrian; **B.** *Paradoxides hicksii* Salter, Middle Cambrian; **C.** *Ctenopyge flagellifera* (Angelin), Dolgelly Beds, Upper Cambrian; **D.** *Peltura scarabaeoides* (Wahlenberg), Dolgelly Beds, Upper Cambrian; **E.** *Dictyonema flabelliforme* (Eichwald), Tremadoc Beds, Upper Cambrian; **F.** *Clonograptus tenellus* (Linnarsson), Tremadoc Beds, Upper Cambrian; **G.** *Asaphellus homfrayi* (Salter), Tremadoc Beds, Upper Cambrian; **H.** *Shumardia pusilla* (Sars), Tremadoc Beds, Upper Cambrian.

the transgression of the sea over a land surface of low relief formed of consolidated and folded Pre-Cambrian rocks. The basal quartzites, laid down in shallow coastal waters, are probably largely residual deposits resulting from prolonged denudation of the windswept land area. The succeeding glauconitic sandstones and phosphatic limestones also indicate shallow water conditions, probably with slow deposition. At the close of Lower Cambrian times minor folding and uplift took place, and the apparent absence of Middle Cambrian strata from the Malverns may indicate that this area remained above sea-level until the Upper Cambrian. Sedimentation continued intermittently in Shropshire with the deposition of shales, glauconitic sandstones and thin breccia beds. The fine-grained Upper Cambrian shales were laid down in quiet water conditions, with breaks in deposition at the base of the Upper Cambrian and Tremadoc Series. Subsidence was more prolonged during Tremadoc Series time with the deposition of a considerable thickness of shales. The occurrence of sandstones in the higher beds in Shropshire suggests a progressive shallowing of the sea from the west towards the end of the Cambrian Period.

The Cambrian rocks in Shropshire and the Malvern area are classified as follows:

### Shropshire

|  | | *Approximate thickness* | |
|---|---|---|---|
|  | | *feet* | *(metres)* |
| Upper Cambrian | | | |
| Tremadoc Series—Shineton Shales . . . | | ? over 3000 | (900) |
| Dolgelly Stage {Black Shales. . . . | | 15 | (4·6) |
| {Grey (*Orusia*) Shales . . | | 65 | (19·8) |
| Middle Cambrian | | | |
| Upper Comley Series . . . . . | | 300 to ?over 600 | (90–180) |
| Lower Cambrian | | | |
| Lower Comley Series {Lower Comley Limestones | | up to 6 | (1·8) |
| {Lower Comley Sandstone | | 500 | (150) |
| Wrekin Quartzite . . . . . . | | up to 150 | (46) |

### Malvern Area

| | feet | (metres) |
|---|---|---|
| Upper Cambrian | | |
| Tremadoc Series—Bronsil Shales . . . | 1000 | (300) |
| Maentwrog (?) and Dolgelly Stages— | | |
| Whiteleaved Oak Shales . . . . | 500 | (150) |
| Middle Cambrian (probably not represented) | | |
| Lower Cambrian | | |
| Hollybush Sandstone . . . . . . | ? over 900 | (270) |
| Malvern Quartzite . . . . . . | ? 300 | (90) |

## Lower Cambrian

### Shropshire

The basal division of the Lower Cambrian is the Wrekin Quartzite, which rests with strong unconformity on the Pre-Cambrian. It is normally a white or pale grey fine-grained quartzite (Plate VA), locally saccharoidal, usually consisting of rounded quartz grains cemented by secondary silica. On the flanks of the Wrekin it is well exposed to a thickness of about 150 ft (46 m), and in its lower beds contains many fragments of the underlying Uriconian rhyolites and tuffs. To the west it thins to about 90 ft (27 m) in the Rushton Anticline, where it rests on the Rushton Schists, and to about 60 ft (18 m) in the Charlton Hill Syncline. At Rushton the basal beds of the Quartzite consist of thin alternations of quartzite with dark grey shale and red-stained sandstone. Farther south the Wrekin Quartzite rests on the Uriconian of the Lawley and Caer Caradoc and the Western Longmyndian of Cwms. Its thickness is very variable, locally reaching 140 ft (43 m), and it is sometimes pebbly near the base. At Hill End, south of Cardington, about 60 ft (18 m) of white quartzite is seen, overlying Uriconian quartz-porphyry.

The basal quartzite has not been detected in the Cambrian areas associated with the Western Uriconian rocks.

In the Wrekin–Caradoc district the Quartzite is succeeded by the Lower Comley Sandstone. Thin beds of conglomerate and greenish grey sandstone at the base contain the oldest known fauna in the district, including small horny brachiopods such as *Paterina phillipsi* and *Obolella? groomi*, and species of *Hyolithellus*. The main part of this division is a greenish glauconitic micaceous sandstone with some shale beds in the Wrekin area. A Geological Survey borehole at Shootrough, near Comley (Fig. 16) proved about 420 ft (128 m) (true thickness) of Lower Comley Sandstone overlain unconformably by Middle Cambrian. The upper part of the sandstone was medium-grained and glauconitic, the lower part fine-grained and silty. The total thickness of the division is probably about 500 ft (150 m). Fossils are not common in the sandstone, but they include various species of horny brachiopods, fragments of trilobites including *Holmia*, ostracods and hyolithids.

Glauconitic sandstone, probably of Lower Cambrian age, was proved beneath Ruabon Marl in a borehole put down from the bottom of the Cruckmeole Shaft at Hanwood Colliery (Fig. 43).

A group of highly fossiliferous sandy limestones, the Lower Comley Limestones, totalling about 6 ft (1·8 m) in thickness, occurs above the Lower Comley Sandstone at the top of the Lower Cambrian in the Wrekin–Caradoc district. From an exposure of these beds in Comley Quarry [SO 4845 9647], Lapworth, in 1888, recorded the first Olenellid trilobite to be found in Britain. Between 1906 and 1936 Cobbold studied the fauna of these sandy limestones and he divided them into beds characterized in ascending order by the trilobites *Callavia* [*Olenellus*], '*Eodiscus*' *bellimarginatus*, *Strenuella* and *Protolenus*, and the pteropod-like shell *Lapworthella nigra*. The section at Comley Quarry is:

|  |  | *ft* | *in* | *(cm)* |
|---|---|---|---|---|
| *Lapworthella* Limestone |  |  |  |  |
| Very dark grey, composed of phosphatic material and occasional quartz pebbles. Locally thinning out | up to | – | 6 | (15) |
| *Protolenus* Limestone |  |  |  |  |
| Pale grey fossiliferous limestone, dark and phosphatic where fossils are rarer | about | – | 6 | (15) |
| *Strenuella* Limestone |  |  |  |  |
| Red to purple sandy limestone with well-rounded grains of quartz and a phosphatic matrix | about | – | 9 | (23) |

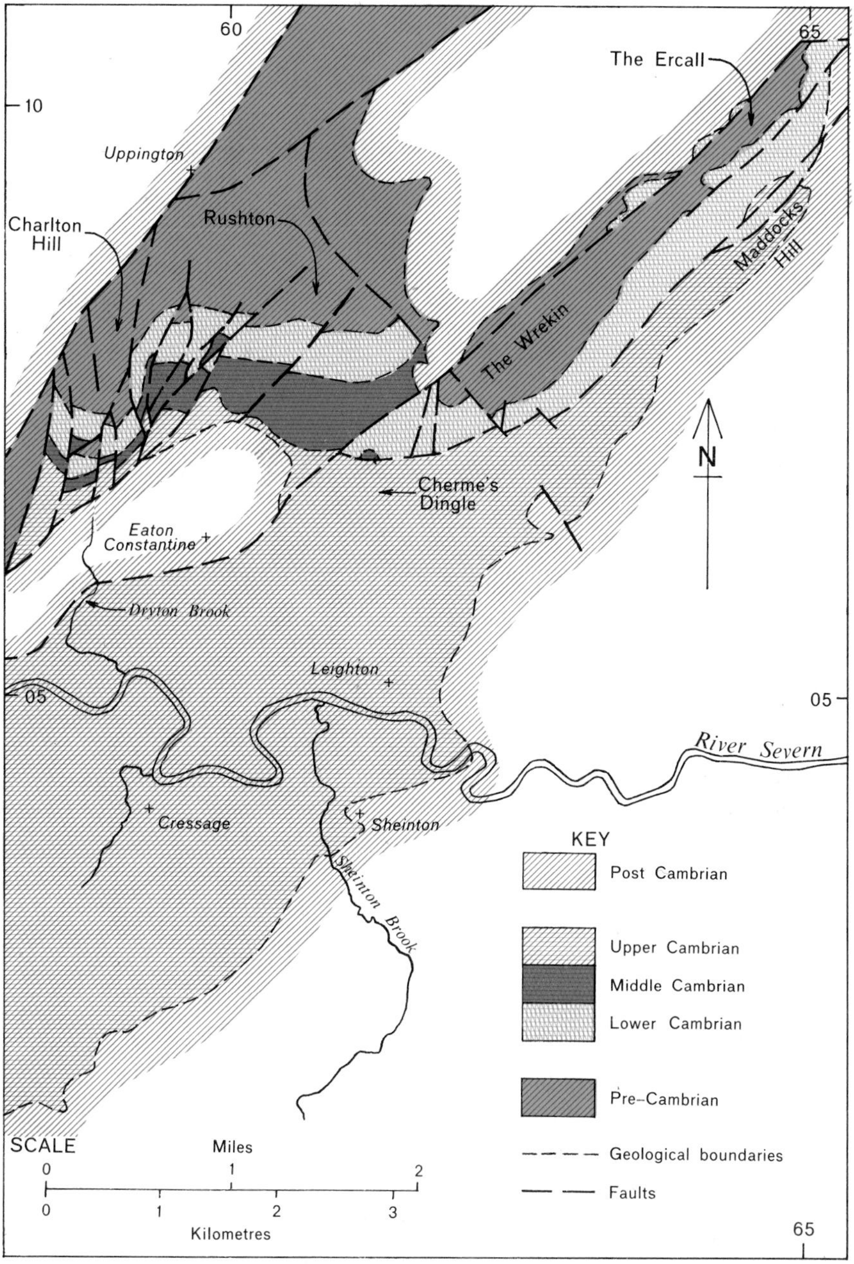
60
65
10
05
The Ercall
Uppington
Charlton Hill
Rushton
Maddocks Hill
The Wrekin
N
Cherme's Dingle
Eaton Constantine
Dryton Brook
Leighton
River Severn
Cressage
Sheinton
Sheinton Brook
KEY
Post Cambrian
Upper Cambrian
Middle Cambrian
Lower Cambrian
Pre-Cambrian
Geological boundaries
Faults
SCALE
Miles
0
1
2
0
1
2
3
Kilometres
65

FIG. 15. *Sketch-map of the Cambrian rocks of the Wrekin area*

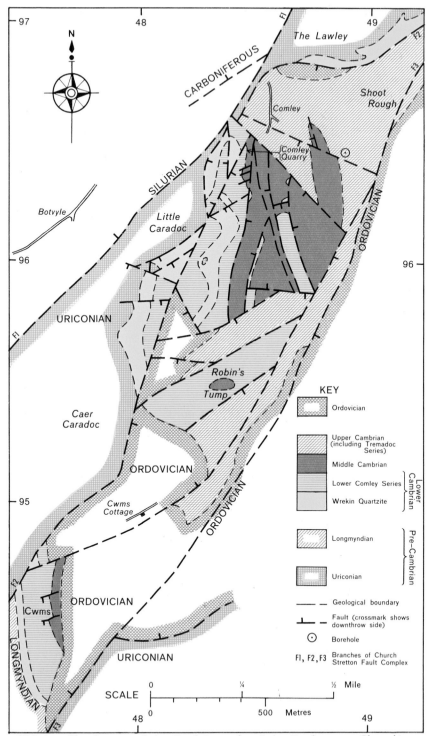

**KEY**

Ordovician

Upper Cambrian
(including Tremadoc
Series)

Middle Cambrian

Lower Comley Series

Wrekin Quartzite

⎱ Lower Cambrian

Longmyndian

Uriconian

⎱ Pre-Cambrian

Geological boundary

Fault (crossmark shows
downthrow side)

Borehole

F1, F2, F3  Branches of Church
Stretton Fault Complex

SCALE

0     ¼     ½  Mile

0     500  Metres

**FIG. 16**  *Sketch–map of the Cambrian rocks of the Comley area, Shropshire*

'*Eodiscus' bellimarginatus* Limestone

Phosphatic limestone . . . . . . about 1 9 (53)

Red *Callavia* Sandstone (Raw 1936) [*Olenellus* Limestone of Lapworth]

Nodules of red or purplish, micaceous and glauconitic calcareous sandstone . . . . about 2 6 (76)

Green *Callavia* Sandstone (Raw 1936)

Bright green glauconitic sandstone . . . thickness not known

In the Shootrough Borehole, about ¼ mile (0·4 km) east of Comley Quarry, the Lower Comley Limestones were absent due to post-Lower Cambrian erosion; they are probably also absent in the Hill End area, south of Cardington. However, a strictly comparable sequence of limestones has been described by Cobbold and Pocock (1934) from the Rushton area west of the Wrekin (Fig. 15).

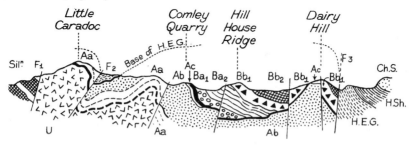

FIG. 17. *Generalized section across the Comley Cambrian area, showing the principal tectonic features*

(Length = about ½ mile, 0·8 km)

The Silurian on the west is part of the All Stretton area, showing the Wenlock and Aymestry limestones.

Caradocian: Ch.S. = Chatwall Flags and Sandstone; H.Sh. = Harnage Shale; H.E.G. = Hoar Edge Grit. Cambrian: Bb2 = *Paradoxides intermedius* Shale; Bb1 = *P. intermedius* Grit and Breccia (Comley Breccia Bed); Ba2 = *P. groomi* Shale; Ba1 = *P. groomi* Grit and Conglomerate; Ac = Lower Comley Limestones (*Lapworthella, Protolenus, Strenuella, 'Eodiscus' bellimarginatus, Callavia*); Ab = Lower Comley Sandstone; Aa = Wrekin Quartzite; U = Uriconian.

Thickened sinuous lines indicate three unconformities: **1.** At the base of the Cambrian (over Little Caradoc); **2.** At the base of the Middle Cambrian (Comley Quarry to Dairy Hill); **3.** At the base of the Caradoc Series (from near F2 to F3).

F1, F2 and F3 are branches of the Church Stretton Fault Complex.

(After Cobbold 1927.)

## Malvern area

The basal Malvern Quartzite is similar in lithology to the Wrekin Quartzite of Shropshire. It contains beds of conglomerate with pebbles which include metamorphic quartzite, rhyolite, andesite and red granophyre suggesting derivation in part from the rocks of the Pre-Cambrian Malvern complex. It is poorly exposed and its thickness cannot be estimated accurately. Groom (1902) considered that it might be several hundred feet thick at the northern end of Midsummer Hill. Unlike the Wrekin Quartzite it contains a few fossils including the brachiopods *Paterina phillipsi* and *Obolella ? groomi*, and some hyolithids.

The Malvern Quartzite is overlain by the Hollybush Sandstone which resembles the Lower Comley Sandstone of Shropshire. The basal beds of the Sandstone, about 75 ft (23 m) thick, comprise flaggy and shaly green glauconitic sandstones with thin calcareous layers. They are overlain by massive green glauconitic sandstones with some grey and black sandstone and quartzite in the lower part. The total thickness of the formation is difficult to estimate, but is probably at least 900 ft (270 m). It contains a fauna of horny brachiopods and hyolithids. Groom (1902) considered that the shaly basal beds of the Hollybush Sandstone possibly correspond to the '*Olenellus*' Beds, and the overlying Zone of *Paradoxides groomi* in the Comley area, and that the main part of the Sandstone may represent the greater part of the Paradoxidian (Middle Cambrian) of other localities. However, Stubblefield (1956) has indicated that there is no evidence of Middle Cambrian faunas in the Malvern area and has placed the whole of the Hollybush Sandstone in the Lower Cambrian.

## Middle Cambrian

### Shropshire

A break in the continuity of deposition occurred between the formation of the Lower Comley Limestones and the overlying Middle Cambrian beds. During this interval the Lower Cambrian strata were consolidated, folded and partly eroded before the lowest Middle Cambrian beds were deposited (Fig. 18). In the latter there are fragments of rock containing a Lower Cambrian fauna enclosed in a matrix of sediment with Middle Cambrian fossils.

The Middle Cambrian deposits consist of alternations of glauconitic sandstones and grits with shales and some conglomerates, totalling some 300 to 600 ft (90–180 m) in thickness. The most characteristic fossils are various species of the trilobite genus *Paradoxides* and the Middle Cambrian sequence can be divided, as in Scandinavia, into a number of zones named after different *Paradoxides* species.

At Comley, the lowest beds are the *Paradoxides groomi* Grits and Conglomerate, about 30 ft (9 m) thick, which comprise coarse pebbly grits, locally conglomeratic, with blocks of Lower Cambrian sandstone and limestone. The fauna of the matrix includes *P. groomi* and *Kootenia* [*Dorypyge*] *lakei*. These grits pass upwards into about 300 ft (90 m) of shales which are in turn overlain by a few feet of sandy flags with *Kootenia*. All these beds are thought to fall within the *P. oelandicus* zones of Scandinavia.

The succeeding *P. intermedius* Grit and Breccia (Comley Breccia Bed), up to 35 ft (10·7 m) thick, consists of blocks of Lower Cambrian limestone and sandstone in a glauconitic sandstone matrix. The matrix contains Middle Cambrian fossils such as *Bailiella cobboldi* and *Paradoxides* and also a number of derived Lower Cambrian forms. Locally, as in the Shootrough Borehole, the Comley Breccia Bed rests directly on Lower Cambrian strata (Fig. 18). The Breccia Bed is overlain by poorly exposed shales and sandstone, possibly 300 ft (90 m) thick, which contain *P. intermedius* and other fossils. These shales are succeeded by the '*P. rugulosus*' Sandstones and *P. davidis*

Grits, each at least 10 ft (3 m) thick. These beds, from the Breccia Bed upwards, may fall within the *P. paradoxissimus* zones of Scandinavia.

The uppermost Middle Cambrian beds at Comley are the *Billingsella* Beds, some 6 ft (1·8 m) of calcareous grey sandstones with an extensive brachiopod fauna including *Obolus linnarssoni* and species of Acrotretids, suggesting a possible correlation with the *P. forchhammeri* Zone.

The Middle Cambrian is poorly exposed around Rushton but a sequence, comparable in part with that at Comley, has been worked out by Cobbold and Pocock (1934).

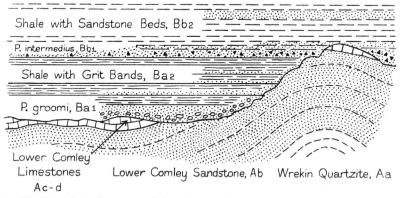

FIG. 18. *Hypothetical section of disturbed and eroded Lower Cambrian beds during the deposition of the Middle Cambrian*

The left-hand end of the section indicates conditions that might have obtained during the accumulation of the *Paradoxides groomi* Grit and Conglomerate (Ba1) of Comley Quarry Ridge; the right-hand end, those that might have given rise to the formation of the *P. intermedius* Grit and Breccia (Comley Breccia Bed) (Bb1) of Comley Brook, Dairy Hill and the Shootrough Borehole. The present horizontal distance apart is from 100 to 200 yd (90–180 m); the thickness of the P. groomi Shales (Ba2) is estimated at 300 ft (90 m).

(From Cobbold 1927.)

# Upper Cambrian

## Shropshire and Herefordshire

Although there is no angular disconformity between the Middle and Upper Cambrian, there appears to be a non-sequence as no representatives of the Maentwrog or Festiniog stages of North Wales have been recognized in the area. The lowest Upper Cambrian beds fall within the Dolgelly Stage and have been described by Stubblefield (1930) from the Bentleyford Brook area, north-east of the Lawley. The sequence begins with about 65 ft (19·8 m) of dark grey micaceous shales (the Grey (*Orusia*) Shales), with some calcareous concretions, which have yielded the brachiopod *Orusia lenticularis*. They are succeeded by the Black Shales, about 15 ft (4·6 m) of black shales with numerous black, bituminous limestone concretions ('stinkstones') aligned along the bedding planes. These concretions, and similar 'stinkstones' in the Dryton Brook (Fig. 15), are of the cone-in-cone type with a homogeneous limestone kernel which in some cases contains trilobites such as *Leptoplastus raphidophorus* and *Ctenopyge flagellifera* (Fig. 14c) indicating the presence of the *Leptoplastus* and lower *Ctenopyge* zones as defined in Sweden. The *Orusia* Shales are present in the Comley area to the south, but it is uncertain if the Black Shales occur there.

There is a further break in deposition between the Black Shales and the overlying Shineton Shales which are correlated with the Tremadoc Series of North Wales. These beds, which have an extensive outcrop south-west of the Wrekin and in the Bentleyford–Comley area, are bluish grey shales with fine-grained sandstone beds in the upper part and with cone-in-cone concretions at many horizons. The thickness of the Shales is not known but Stubblefield and Bulman (1927) estimated it to be greater than 3000 ft (900 m). They have worked out the following sequence (in descending order):

6. Arenaceous Beds [Shales with sandstone beds]
5. Zone of the trilobite *Shumardia pusilla*
4. Brachiopod Beds
3. Zone of the graptolite *Clonograptus tenellus*
2. Transition Beds
1. Zone of the graptolite *Dictyonema flabelliforme*

The three lower divisions can be examined in Cherme's Dingle, south of the Wrekin, in a practically continuous section; while in Shineton Brook on the south side of the Severn Valley a fairly complete sequence is obtained from the Brachiopod Beds to the Zone of *Shumardia pusilla*, which contains many trilobites including *Asaphellus homfrayi* (Fig. 14g). The Arenaceous Beds can be seen farther to the south-west near Evenwood. Shales yielding *Dictyonema flabelliforme* are recorded from the Lawley area and also southeast of Cardington where the Transition Beds are also present.

Farther to the west, in the area between the Long Mynd and the Stiperstones, the Shineton Shales extend in a narrow outcrop from Granham's Moor (Habberley) in the north to Snead (near Lydham) in the south. The lower beds are shales from which *D. flabelliforme* and *Clonograptus tenellus* (Fig. 14f) have been recorded. Arenaceous beds appear to come in at a lower horizon than in the Wrekin area, possibly in the upper part of the *C. tenellus* Zone.

A faulted inlier at Pedwardine near Brampton Bryan (Herefordshire) contains Shineton Shales of the *D. flabelliforme* Zone.

### Malvern area

The Upper Cambrian of the Malvern area comprises the Black (Whiteleaved Oak) Shales, about 500 ft (150 m) thick, overlain by the Grey (Bronsil) Shales which are about 1000 ft (300 m) thick.

The basal beds of the Whiteleaved Oak Shales, immediately above the Hollybush Sandstone, consist of black shales interbedded with coarse very dark grey glauconitic grits. They contain a fauna including the ostracod-like crustacean *Cyclotron* [*Polyphyma*] *lapworthi* and may possibly fall within the Maentwrog Stage. The higher beds are soft black shales which yield numerous fossils especially trilobites including *Peltura scarabaeoides* (Fig. 14d) *Sphaerophthalmus humilis*, *S. major* and several species of *Ctenopyge*, which indicate the *P. scarabaeoides* Zone (Upper Dolgelly Stage).

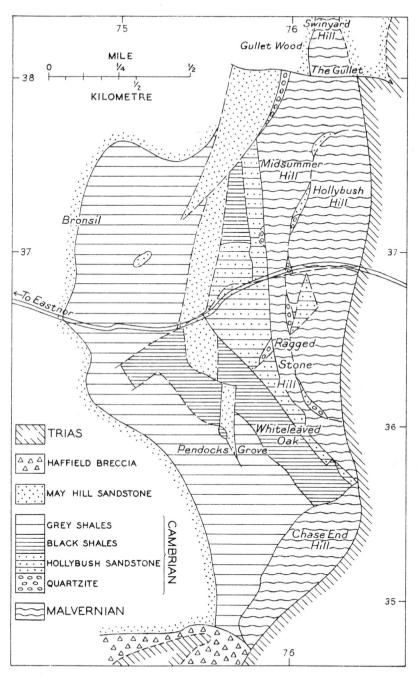

FIG. 19. *The Cambrian area of the southern Malverns*
(After Groom 1899.)

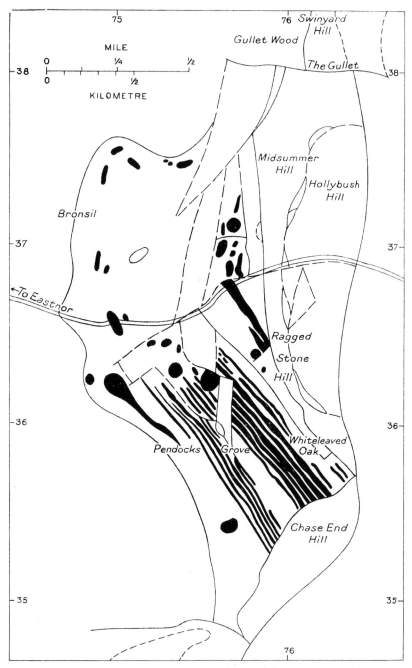

Fig. 20. *Igneous intrusions in the Cambrian of the southern Malverns*
(After Groom 1899.)

The Bronsil Shales occupy a considerable area west of the southern Malverns. They are usually grey shales with a light blue, light green or yellowish tint and contain *Dictyonema flabelliforme*, brachiopods and trilobites including *Euloma monile* and *Shumardia*. They are equivalent to the lower part of the Shineton Shales (Tremadoc Series).

## Igneous Intrusions in the Cambrian

There do not appear to be any contemporaneous igneous rocks in the Cambrian beds of the Welsh Borderland, but intrusions of later date, probably late Ordovician, occur in the Shineton Shales at Maddocks Hill east of the Wrekin and near Eastridge in the Pontesbury district, and in various strata in the Malvern area.

The Maddocks Hill rock occurs as a dyke or sill in the nearly vertical Shineton Shales and crops out over a distance of $\frac{1}{2}$ mile (0·8 km) with a maximum width of 90 yd (82 m). The shales are baked and otherwise altered by the intrusion for some 25 yd (23 m) from the contact. The rock, which has been described as a camptonite or albite-diabase, is of fine-grained crystalline type and consists mainly of feldspar, augite, decomposed olivine and hornblende. Its mottled appearance is due to the contrast between the pink feldspar crystals and the dark green colour of the other minerals. A similar type is intrusive in the Cambrian rocks of the Malvern district, at Nuneaton, in Leicestershire and in other Cambrian areas of England and Scotland.

In the Pontesbury district there are small dykes and sills of dolerite and quartz-dolerite, similar in character to the intrusions in the Longmyndian (p. 25).

In the Malvern area there are, beside the camptonite type (also described as spilitic andesites), olivine-bearing basalts (or amygdaloidal spilites) and diabases which penetrate both the Grey and Black Shales and, more rarely, the Hollybush Sandstone (Fig. 20).

## References

BLYTH, F. G. H. 1935. The basic intrusive rocks associated with the Cambrian inlier near Malvern. *Quart. J. Geol. Soc. Lond.*, **91**, 463–78.

CALLAWAY, C. 1877. On a new Area of Upper Cambrian Rocks in South Shropshire, with a description of a new Fauna. *Quart. J. Geol. Soc. Lond.*, **33**, 652–72.

—— 1878. On the Quartzites of Shropshire. *Quart. J. Geol. Soc. Lond.*, **34**, 754–63.

COBBOLD, E. S. 1921. The Cambrian Horizons of Comley (Shropshire) and their Brachiopoda, Pteropoda, Gastropoda, etc. *Quart. J. Geol. Soc. Lond.*, **76** (for 1920), 325–86.

—— 1927. The Stratigraphy and Geological structure of the Cambrian Area of Comley (Shropshire). *Quart. J. Geol. Soc. Lond.*, **83**, 551–73.

—— and POCOCK, R. W. 1934. The Cambrian area of Rushton (Shropshire). *Phil. Trans. Roy. Soc.* (B), **223**, 305–409.

COX, A. H. 1912. On an Inlier of Longmyndian and Cambrian Rocks at Pedwardine (Herefordshire). *Quart. J. Geol. Soc. Lond.*, **68**, 364–73.

GREIG, D. C., WRIGHT, J. E., HAINS, B. A. and MITCHELL, G. H. 1968. Geology of the country around Church Stretton, Craven Arms, Wenlock Edge and Brown Clee. *Mem. Geol. Surv.*

GROOM, T. T. 1899. The Geological Structure of the Southern Malverns, and of the Adjacent District to the West. *Quart. J. Geol. Soc. Lond.*, **55**, 129–69.

—— 1901. On the Igneous Rocks associated with the Cambrian Beds of the Malvern Hills. *Quart. J. Geol. Soc. Lond.*, **57**, 156–84.

—— 1902. The Sequence of the Cambrian and Associated Beds of the Malvern Hills. *Quart. J. Geol. Soc. Lond.*, **58**, 89–135.

—— 1910. The Malvern and Abberley Hills and the Ledbury District. *In* Geology in the Field (Jubilee Volume of the Geologists' Association), 698–738.

LAKE, P. 1906–46. A Monograph of the British Cambrian Trilobites. *Palaeont. Soc.* [*Monogr.*].

LAPWORTH, C. 1888. On the Discovery of the *Olenellus* Fauna in the Lower Cambrian Rocks of Britain. *Geol. Mag.* (3), **5**, 484–7.

POCOCK, R. W., WHITEHEAD, T. H., WEDD, C. B. and ROBERTSON, T. 1938. Shrewsbury District including the Hanwood Coalfield. *Mem. Geol. Surv.*

RAW, F. 1936. Mesonacidae of Comley in Shropshire, with a discussion of classification within the family. *Quart. J. Geol. Soc. Lond.*, **92**, 236–93.

STUBBLEFIELD, C. J. 1930. A new Upper Cambrian section in south Shropshire. *Sum. Prog. Geol. Surv.* for 1929, **2**, 54–62.

—— 1956. Cambrian Palaeogeography in Britain. *Rep. XX Int. Geol. Congr.* (Mexico), Symposium I, 1–43.

—— and BULMAN, O. M. B. 1927. The Shineton Shales of the Wrekin District: with Notes on their Development in other parts of Shropshire and Herefordshire. *Quart. J. Geol. Soc. Lond.*, **83**, 96–146.

# 4. Ordovician

Ordovician rocks are present at the surface only in the northern part of the Welsh Borderland—in Shropshire, the adjacent parts of Montgomeryshire, and at Pedwardine in Herefordshire. Farther south they are probably absent, since, near the Malvern Hills and elsewhere, Silurian strata rest directly on Cambrian and Pre-Cambrian rocks.

The principal areas of Ordovician rocks are the Breidden Hills, the Stiperstones, Shelve and Chirbury area, the Pontesford area and the 'Caradoc' area (between Harnage and Coston). Smaller outcrops occur at Forden, near Welshpool, and at Pedwardine.

During Ordovician times the Welsh Borderland lay between the subsiding Welsh Geosyncline to the west and the relatively stable Midland Block to the east. This led to profound variations in sedimentation within the area. At the close of the Cambrian Period the whole area was raised above sea-level and denudation and minor folding took place. Subsidence recommenced during the Arenig epoch in the western part of the Welsh Borderland (Shelve area) and continued, apparently with little or no break, until at least Caradoc times. A thick sequence, largely of graptolitic shales, was laid down. Farther east, the Pontesford and 'Caradoc' areas lay within the influence of the Midland Block. Here, the earliest Ordovician sediments are of Caradoc age and are mainly of shallow water shelly facies, in contrast with the graptolitic shales in the west. Still farther east, in the Midlands, Ordovician rocks are unknown (Fig. 21). Considerable denudation of the Ordovician rocks took place in early Silurian times so that the basal Silurian strata rest with strong unconformity on various divisions of the Ordovician, or on earlier rocks. The highest Ordovician rocks (Ashgill Series), if present, are everywhere concealed, although they occur in North Wales.

Many of the Ordovician rocks are fossiliferous (Figs. 22, 23). In the western areas of outcrop they consist largely of shales and flags, and in such beds graptolites are not uncommon. The strata have been divided into zones, each characterized by the association of certain species and each named after a prominent species. Similarly, in the eastern areas, a sequence of stages based on the shelly faunas (mainly brachiopods and trilobites) has been worked out. It has proved difficult to correlate the two facies, but recent work by Whittard and Dean has to a large degree solved this problem (Fig. 31).

## Stiperstones, Shelve and Chirbury area

In this area (see map, Plate IV) the sequence of the Ordovician rocks seems to be almost complete, apart from the absence of the upper part of the Caradoc Series and the Ashgill Series. Except for the lowest beds (Stiperstones Quartzite) which indicate the proximity of a shoreline to the east, the strata were probably deposited in quiet and moderately deep water, with occasional

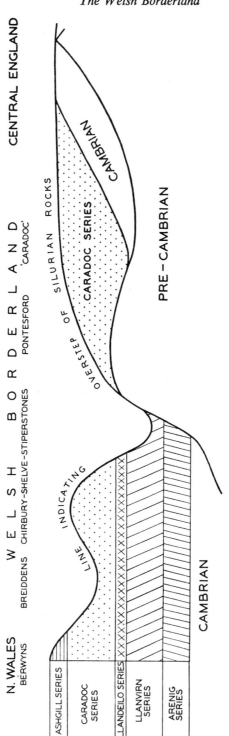

FIG. 21. *Diagram illustrating the relation of the Ordovician of the Welsh Borderland to that in North Wales, where the Ordovician sequence is complete, and to Central England, where these rocks are absent*

influxes of coarser sediment and outbursts of activity from volcanic or submarine craters.

The Ordovician strata of this area were subdivided and mapped by Lapworth. More recently, Whittard has remapped the area and has made an extensive study of the trilobite faunas. His maps are, as yet, unpublished. The igneous rocks have been studied by Watts and by Blyth. The sequence of strata is as follows:

|  |  | Approximate thickness feet (metres) | |
|---|---|---|---|
| Caradoc Series | Whittery Shales . . at least | 1000 | (300) |
| | Whittery Volcanic Group | 300 | (90) |
| | Hagley Shales . . | ?150 | (45) |
| | Hagley Volcanic Group | 350 | (100) |
| | Aldress Shales . . | 1000 | (300) |
| | Spy Wood Grit . | 300 | (90) |
| | Rorrington Beds . | 400 | (120) |
| Llandeilo Series | Meadowtown Beds . | 400 | (120) |
| Llanvirn Series | Betton Beds . . | 600 | (180) |
| | Weston Beds . . | | |
| | Stapeley Shales . . | 5000 | (1500) |
| | Stapeley Volcanic Group | | |
| | Hope Shales . . | | |
| Arenig Series | Tankerville Flags | | |
| | Mytton Flags . at least | 3000 | (900) |
| | Stiperstones Quartzite | | |

The thicknesses in this table are mainly from Whittard (1952); they differ in several respects from those of Lapworth on Plate IV and Fig. 24.

### Arenig Series

The Stiperstones Quartzite at the base of the Arenig Series rests unconformably on Cambrian (Tremadoc Series) Shales. It is a hard white or grey siliceous sandstone with beds of conglomerate, and with occasional thin shale beds. It forms the long ridge of the Stiperstones (Plate IIIB) and is well exposed in crags on this ridge such as the Devil's Chair and Cranberry Rock and also in quarries near Pontesbury. Traces of boring organisms occur, and also the rare trilobite *Neseuretus grandior*.

The overlying Mytton Flags occur in two outcrops, one to the west of the Stiperstones ridge and the other in the dome of Shelve Hill. They comprise bluish grey shales and flags which are well exposed west of the Stiperstones and which also form most of the spoil heaps of the old lead mines near Shelve. There is a mixed shelly and graptolitic fauna with horny brachiopods such as *Monobolina plumbea* (Fig. 22C) and trilobites including *Ogygiocaris selwynii* and the early trinucleid *Myttonia multiplex*. Extensiform graptolites (in which the stipes are in line), including the zonal form *Didymograptus extensus* (Fig. 22A), occur in the higher beds; brown weathering micaceous flags at Shelve church contain many graptolites.

The Tankerville Flags, which succeed the Mytton Flags, have a distinct trilobite fauna and contain the zonal graptolite *Didymograptus hirundo* (Fig. 22B).

### Llanvirn Series

This Series begins with the Hope Shales, a thick sequence of rusty weathering black shales with beds of very fine-grained volcanic tuff ('chinastone ash') in the middle and upper part. Almost the whole thickness of the Shales can be seen in the Hope Valley, and the chinastone ash is well exposed near Hope Rectory. The Hope Shales contain an extensive trilobite fauna; among the commonest forms are *Barrandia homfrayi* and *Stapeleyella inconstans* while many other less common species (e.g. *Ectillaenus hughesi* Fig. 22H) are restricted to the Shales. Graptolites include the zonal form *Didymograptus bifidus* (Fig. 22E).

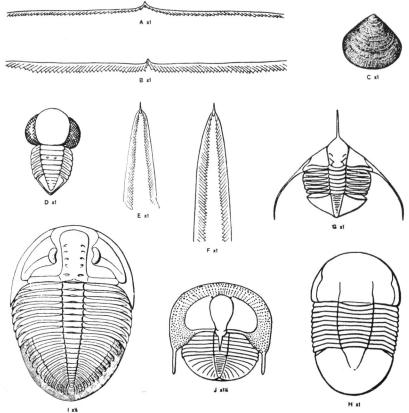

FIG. 22. *Ordovician Fossils* (*Arenig, Llanvirn and Llandeilo Series*)
All figures natural size except I × ½ and J × 1½

Arenig Series: **A.** *Didymograptus extensus* (Hall), Mytton Flags; **B.** *Didymograptus hirundo* Salter, Tankerville Flags; **C.** *Monobolina plumbea* Salter, Mytton Flags.

Llanvirn Series: **D.** *Pricyclopyge binodosa* (Salter), Hope Shales; **E.** *Didymograptus bifidus* (Hall), Hope and Stapeley Shales; **F.** *Didymograptus murchisoni* (Beck), Betton Beds; **G.** *Ampyx linleyensis* Whittard, Hope Shales and Stapeley Volcanic Group; **H.** *Ectillaenus hughesii* (Hicks), Hope Shales.

Llandeilo Series: **I.** *Ogygiocarella debuchii* (Brongniart), Meadowtown Beds; **J.** *Lloydolithus lloydii* (Murchison), Meadowtown Beds.

(A11100)

A. Hanter Hill, Worsel Wood and Stanner Rocks

*(For full explanation see p. ix)*

**Plate III**

B. Northward view along the Stiperstones, Shropshire

(A11101)

Above the Hope Shales is the Stapeley Volcanic Group in which waterlain andesitic tuffs and breccias with some lavas are interbedded with shales. Stapeley Hill is formed of this Group, and a well-known section in it is that of Tasgar Quarry, adjacent to the main road about 1¼ miles (2 km) north-north-east of Hyssington. The shales within the Stapeley Volcanic Group and the succeeding bluish grey Stapeley Shales yield a trilobite fauna with many similarities to that of the Hope Shales. Brachiopods and graptolites also occur, and the latter indicate that, like the Hope Shales, the Stapeley Volcanic Group and Stapeley Shales fall within the *D. bifidus* Zone.

The Weston Beds are much more arenaceous than the underlying Stapeley Shales. They comprise two groups of massive flags and argillaceous sandstone (Upper and Lower Weston Grits) separated by a group of siltstones and shales (Weston Shales). The shaly beds and interbedded tuffs contain an extensive fauna including trilobites, gastropods, bivalves and horny brachiopods. The trilobite fauna, which includes *Bettonia frontalis* and *Ogyginus corndensis*, differs considerably from that of the Stapeley Shales, and the infrequent graptolites indicate that the beds should be placed in the *Didymograptus murchisoni* Zone (Fig. 22F). Exposures of the Weston Beds can be seen at Priestweston, about 2 miles (3·2 km) south-east of Chirbury.

There is a gradation from the Weston Beds into the overlying Betton Beds which comprise bluish black shales and flags. The trilobite fauna includes *O. corndensis* and *Trinucleus acutofinalis*, the latter being restricted to these beds. Graptolites include the zonal form *D. murchisoni*, and *Glyptograptus dentatus*.

### Llandeilo Series

The Llandeilo Series is represented by the Meadowtown Beds, which consist of shales and mudstones with some thin limestones and calcareous tuffs. These beds are well known for their fauna of trilobites, especially trinucleids such as *Cryptolithus inopinatus* and *Lloydolithus lloydi* (Fig. 22J) and asaphids such as *Ogygiocarella debuchii* [*Ogygia buchii*] (Fig. 22I). The latter species occurs in great abundance in the upper part of the sequence. Brachiopods and bivalves are also present, and the graptolites, which include *Diplograptus foliaceus*, are indicative of the *Glyptograptus teretiusculus* Zone.

### Caradoc Series

The lowest division of the Caradoc Series, the Rorrington Beds, comprises sooty bluish mudstones with a predominantly graptolitic fauna. The graptolites in the lower part of the sequence are mainly slender branched uniserial types, including the zonal form *Nemagraptus gracilis* (Fig. 23A). Higher beds contain mainly unbranched and biserial graptolites (e.g. *Climacograptus* and *Orthograptus*) and these beds are placed in the *Diplograptus multidens* Zone as are all the higher divisions of the Caradoc Series in this area (Fig. 31). Trilobites are less abundant than in the Meadowtown Beds. *O. debuchii* is still quite common, and other forms include *Spirantyx calvarina* and several trinucleids.

There is a rapid upward passage from the graptolitic Rorrington Beds to the flaggy calcareous sandstone of the overlying Spy Wood Grit with its

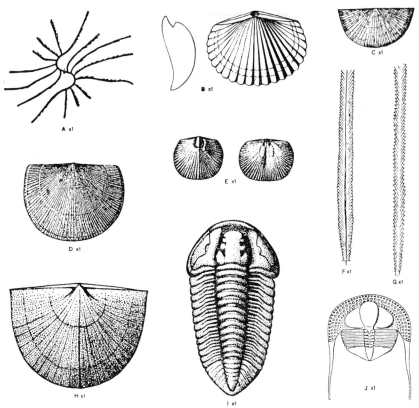

FIG. 23. *Ordovician Fossils (Caradoc Series)*
All figures natural size

**A.** *Nemagraptus gracilis* (Hall), Rorrington Beds and Hoar Edge Grit; **B.** *Dinorthis flabellulum* (J. de C. Sowerby), Hoar Edge Grit; **C.** *Sowerbyella sericea* (J. de C. Sowerby), Lower Cheney Longville Flags; **D.** *Heterorthis alternata* (J. de C. Sowerby), Lower Cheney Longville Flags; **E.** *Dalmanella horderleyensis* (Whittington), Chatwall Sandstone; **F.** *Diplograptus multidens* (Elles and Wood), Pontesford Shales; **G.** *Orthograptus truncatus* (Lapworth), Harnage Shales; **H.** *Kjaerina bipartita* (Salter), Lower Cheney Longville Flags; **I.** *Flexicalymene caractaci* (Salter), Upper Cheney Longville Flags; **J.** *Onnia superba* (Bancroft), Onny Shales.

shelly fauna. This fauna includes an abundance of the ostracod *Tetradella complicata*, brachiopods, starfish and several species of the trilobite genus *Marrolithus*. Thin shaly beds occur within the Grit; these slowly increase in importance upwards and there is a gradual passage into the brown-weathering shales of the Aldress Shales. The graptolite fauna of the Shales includes *Orthograptus truncatus* and *Dictyonema fluitans*. Trilobites are not common, but the forms *Broeggerolithus broeggeri* and *Salterolithus caractaci* appear in the middle of the Shales and range up to the middle of the Whittery Shales indicating that all these higher divisions of the Caradoc Series can be correlated with the Lower Soudleyan Stage of the Caradoc area (Fig. 31).

Above the Aldress Shales there are two volcanic groups, the Hagley and Whittery Volcanic Groups, separated by the bluish black Hagley Shales. The volcanic groups contain massive crystal and lithic tuffs, often brecciated and agglomeratic, and occasional lavas. Fossils found in the tuffs include sponge spicules and graptolites. A good exposure of the Hagley Volcanic Group is afforded by a roadside quarry at Hagley, while the Whittery Volcanic Group is exposed in quarries in the woods about 1 mile (1·6 km) south-east of Chirbury. The highest Ordovician strata in the Shelve area are the Whittery Shales, soft bluish black rusty-weathering shales with rare graptolites, trilobites and other fossils.

The Ordovician rocks of the Shelve area are affected by a number of folds and faults. The main folds are a north-north-easterly trending syncline through Venusbank, Ritton Castle and Pultheley, and a complementary anticline through Hope, Shelve and Hyssington. In the deeper part of the Ritton Castle Syncline beds up to the Stapeley Shales are brought in, while the Shelve Anticline brings up an inlier of the Mytton Flags. According to Whittard (1952, p. 156) the area is complicated by many tear faults, and it is evident that the structure is more complex than is indicated by the section (Fig. 24) and map (Plate IV) reproduced here.

### Igneous intrusions

Numerous igneous intrusions occur within the Stiperstones, Shelve and Chirbury area (Blyth 1944). The largest mass is the phacolith at Corndon Hill, where quartz-dolerite is intruded into a pre-existing anticlinal axis associated with the Shelve Anticline (Fig. 24). Farther south, at Squilver Hill, there is a thick sill of coarse-grained dolerite which is apparently overlain unconformably by Silurian (Upper Llandovery) rocks. Altered dolerites form several sills on Stapeley Hill and to the east of Betton, and there is a group of east to west trending dykes of ophitic dolerite between Betton and Stanage Coppice, about 3 miles (4·8 km) to the east (Plate IV). Many of the dolerite intrusions bear a close resemblance to those in the Cambrian and Longmyndian sediments.

At Cwm Mawr, about 1 mile (1·6 km) south of Corndon Hill, there is an intrusion of picrite, an ultrabasic rock composed mainly of olivine and pyroxene.

Augite-andesite intrusions form the hills of Todleth, Roundton and Lan Fawr, near Church Stoke. They cut across the bedded rocks of Llanvirn age, but their composition is so similar to that of the lavas in the Stapeley Volcanic Group that they are probably not of greatly different age.

Small intrusions of syenitic aplite (quartz-bostonite) occur near Snead south of Hyssington. They appear to be intrusive into Silurian rocks, and resemble veins which cut the Corndon Hill dolerite.

All the igneous intrusions belong to a suite with affinities to an alkaline plateau-basalt or olivine-basalt magma type. Their intrusion, except perhaps for the late-stage aplites, was probably associated with the pre-Upper Llandovery folding of the Ordovician rocks.

## Breidden Hills

The Ordovician inlier of the Breidden Hills area (Figs. 25, 26) is located on the southern side of the River Severn about 12 miles (19 km) west of Shrews-

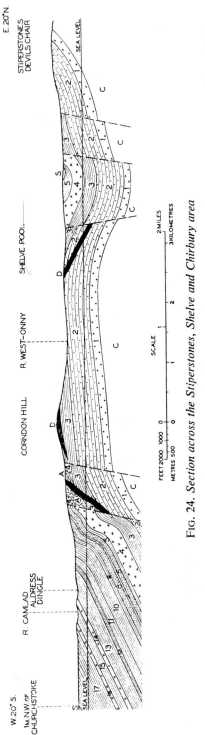

FIG. 24. *Section across the Stiperstones, Shelve and Chirbury area*

**1.** Stiperstones Quartzite; **2.** Mytton and Tankerville Flags; **3.** Hope Shales; **4.** Stapeley Volcanic Group; **5.** Stapeley Shales; **6, 7, 8.** Weston Beds (**6.** Lower Weston Grit; **7.** Weston Shales; **8.** Upper Weston Grit); **9.** Betton Beds; **10.** Meadowtown Beds; **11.** Rorrington Beds; **12.** Spy Wood Grit; **13.** Aldress Shales; **14.** Hagley Volcanic Group; **15.** Hagley Shales; **16.** Whittery Volcanic Group; **17.** Whittery Shales; C. Cambrian; S. Silurian; D. Dolerite; A. Andesite.

(Based upon map by C. Lapworth.)

bury. This inlier, which is about 6 miles (9·7 km) long and up to 1½ miles (2·4 km) wide, is probably limited to the north-west by a fault under the alluvium of the Severn; to the south-east the Ordovician rocks are overlain unconformably by the basal Silurian.

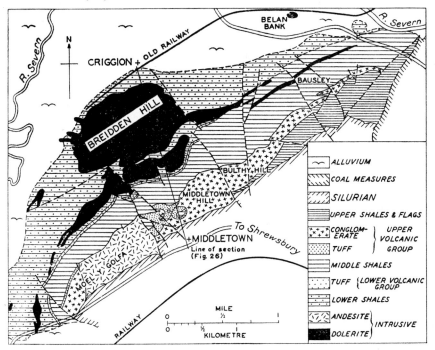

FIG. 25. *Map of the Ordovician rocks of the Breidden Hills*
(After Watts 1925; Wedd and others 1932.)

The Ordovician strata are at least 1500 ft (450 m) thick and have been divided into the following groups, in ascending order: Lower Shales, Lower Volcanic Group, Middle Shales, Upper Volcanic Group, Upper Shales. All these beds fall within the lower part of the Caradoc Series. The Lower Shales can probably be placed in the *Nemagraptus gracilis* Zone and the remainder in the *Diplograptus multidens* Zone. The Lower Shales are folded into an east-north-easterly trending anticline with some associated minor folds. Higher beds are only seen on the south-south-easterly limb of the fold (Fig. 26). This anticline has been intruded by a dolerite laccolith which forms the prominent steep-sided Breidden Hill.

The lowest exposed beds, the Lower Shales, are dark grey shales with graptolites including *Nemagraptus sp.* They are succeeded by the Lower Volcanic Group which consists of tuffs, locally of the fine-grained 'chinastone' type, associated with a black grit containing sponge spicules. This group is developed mainly in the western and middle part of the range, but is represented by hard splintery tuff in the bank of the River Severn, north-east of Bausley. The overlying Middle Shales contain a number of graptolites including *D. multidens* (Fig. 23F).

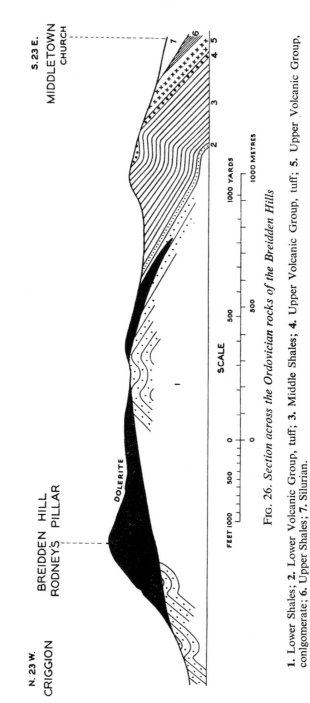

FIG. 26. *Section across the Ordovician rocks of the Breidden Hills*

1. Lower Shales; **2.** Lower Volcanic Group, tuff; **3.** Middle Shales; **4.** Upper Volcanic Group, tuff; **5.** Upper Volcanic Group, conlgomerate; **6.** Upper Shales; **7.** Silurian.

(Mainly after Watts 1925.)

At the base of the succeeding Upper Volcanic Group there are white siliceous tuffs and chinastone which occur at the southern end of Moel y Golfa and above Middletown but disappear northwards. They are followed by a 'bomb-rock', a coarse conglomerate with rounded boulders (or bombs) of andesite in a tuffaceous matrix. At the north-eastern end of its outcrop this conglomerate is overlain by bedded tuffs with graptolites and other fossils. The topmost group, the Upper Shales, comprises grey shales with silty flags and, in places, thin limestones. These shales contain an extensive fauna of graptolites, including *D. multidens*, brachiopods, trilobites and ostracods.

The intrusive dolerite mentioned earlier resembles that of Corndon Hill (p. 45). In addition the Ordovician rocks are intruded by a sill of andesite which forms the crest and south-east flank of Moel-y-Golfa. This sill is overlain unconformably by Silurian (Llandovery Series) rocks.

Ordovician rocks form another small inlier around Forden, south of Welshpool. The rocks are poorly exposed and have not been studied in detail.

## Pontesford area

Ordovician rocks, the Pontesford Shales, crop out to the west of the Longmyndian near Pontesford (Plate VB), and are exposed in the Habberley Brook. They were originally mapped as one continuous outcrop, but a recent interpretation (Dean and Dineley 1961) suggests that they occur in two distinct areas separated by an outcrop of Uriconian rocks (Fig. 27).

In the northern outcrop they rest unconformably on the Uriconian in the Habberley Brook. The basal bed, up to 13 ft (4 m) thick, is a coarse poorly sorted conglomerate which is overlain by about 15 ft (4·6 m) of pyritous silty shales with graptolites including *Diplograptus multidens* (Fig. 23F). These shales are succeeded by some 150 ft (46 m) of interbedded shales, mudstones and siltstones with graptolites and shelly fossils including the trilobites *Broeggerolithus broeggeri* and *Salterolithus caractaci* at the top. The sequence continues with grey shales of uncertain thickness.

In the southern area, exposure is virtually limited to one section in the brook, east of Earl's Hill. Whitehead (1929) considered that, at this locality, the basal conglomerate of the Pontesford Shales rested on an eroded rhyolite, possibly of Ordovician age, which in turn was unconformable on the Longmyndian. It has more recently been suggested (Dean and Dineley 1961) that this 'rhyolite' is not a continuous sheet but consists of massive boulders of rhyolite, probably of Uriconian derivation, in the lower part of the basal conglomerate.

The Pontesford Shales lie within the *D. multidens* Zone, and the upper shelly fauna can also be correlated with the Lower Soudleyan Stage (upper part of the Harnage Shales) of the Caradoc area (Fig. 31).

## 'Caradoc' Area

This outcrop of Ordovician (Caradoc Series) rocks extends for about 19 miles (31 km) from Harnage (Fig. 28) in the north to Coston in the south, and has a maximum width of about 2 miles (3·2 km). It is divided in two by the Pre-Cambrian and Cambrian of the Cardington area. The Ordovician, which has a general south-easterly dip, rests on Pre-Cambrian and Tremadoc Series rocks and is overlain by the local basal Silurian (Llandovery Series). Only the Caradoc Series is present in this area, and the rocks are mainly shallow-water deposits with an abundant shelly fauna of brachiopods and

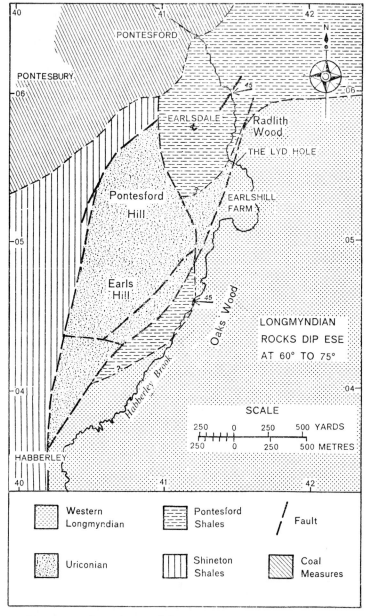

FIG. 27. *Geological map of the Pontesford district of Shropshire*
(From Dean and Dineley 1961.)

trilobites. This contrasts with the Shelve area (p. 43) to the west where deeper water deposits contain a mixed graptolitic and shelly fauna. Bancroft divided the Caradoc Series into stages (Fig. 31) on the basis of successive brachiopod–trilobite faunas, and this work was later amplified by Dean (1958) who also worked out a correlation between the graptolitic and shelly faunas.

The sequence is as follows:

|  | *Approximate thickness* | |
|---|---|---|
|  | *feet* | (*metres*) |
| Onny Shales . . . . . . . | up to 400 | (120) |
| Acton Scott Group . . . . . . | 200 to 500 | (60–150) |
| Cheney Longville Flags (including the *alternata* Limestone) . . . . . . . | 350 to 850 | (110–260) |
| Chatwall Sandstone . . . . . . | 120 to 500 | (37–150) |
| Chatwall Flags . . . . . . | 100 to 300 | (30–90) |
| Harnage Shales . . . . . . . | 300 to 1000 | (90–300) |
| Hoar Edge Grit . . . . . . | 0 to 400 | (0–120) |

*Hoar Edge Grit.* This, the basal formation of the Caradoc Series, takes its name from the prominent scarp of Hoar Edge (Fig. 28) where it reaches its maximum thickness of 400 ft (120 m). Here it consists largely of sandstones and pebbly sandstones, the wind-faceted pebbles and rounded sand grains indicating derivation from an adjacent arid land area. Farther north the Grit thins rapidly and changes laterally into calcareous sandstones with shelly beds. In this northern part of the Caradoc area the lower beds contain the brachiopod *Harknessella subplicata* and the graptolite *Nemagraptus gracilis* (Fig. 23A) and higher beds yield the brachiopod *Harknessella subquadrata*. In the extreme south, at Coston, there are basal quartz conglomerates overlain by sandstones which are often shaly and calcareous. In the Onny Valley the conglomerates are not developed, and farther north between Brokenstones and Cardington Hill the Grit is absent. The fauna in the sandstones includes the brachiopods *Dinorthis flabellulum* (Fig. 23B) and *Harknessella vespertilio* and the trilobite *Costonia ultima*.

In the past (e.g. Pocock and others 1938, and previous editions of this handbook) the overlying Harnage Shales were assigned to the *Dicranograptus clingani* Zone and it was thought that there was a stratigraphical break between the Hoar Edge Grit and Harnage Shales, the *Diplograptus multidens* Zone being absent. However, the Harnage Shales are now placed in the *D. multidens* Zone and there is no evidence of such a break (Fig. 31).

*Harnage Shales.* The Harnage Shales usually occupy a broad hollow between the ridges formed by the Hoar Edge Grit and Chatwall Sandstone, and as a consequence are poorly exposed. North of Cardington Hill they are grey or greenish grey mudstones with some sandstone beds, and may be 1000 ft (300 m) thick. On the southern side of the hill, and westwards to Ragleth Hill they rest unconformably on the Uriconian and there are calcareous pebbly sandstones at the base. Neptunian dykes, with Harnage Shales fossils, occur in the Uriconian of Hazler Hill (p. 18). Over the remainder of the southern Caradoc area the Shales, up to 500 ft (150 m) thick, comprise orange-stained flaggy micaceous silty mudstones which pass upwards into the Chatwall Flags. At Sibdon Carwood there is a thin basaltic lava flow, the only volcanic rock in the Ordovician of the Caradoc area.

The lower part of the Shales (Smeathen Wood Beds of Dean 1958) contains the trilobites *Reuscholithus reuschi* and *Salterolithus caractaci*. The higher beds (Lower Soudleyan Stage, Glenburrell Beds) also contain *S. caractaci*, with *Broeggerolithus broeggeri*, a fauna which is also found in the Shelve area, Pontesford and the Breidden Hills. Brachiopods are quite common and the ostracod *Tallinnella* is often abundant. Graptolites include *Amplexograptus* and *Diplograptus* cf. *multidens*.

*Chatwall Flags.*   The Chatwall Flags, and the succeeding Chatwall Sandstone, take their name from the Chatwall district of Shropshire (Fig. 28). The formation consists of flaggy fine-grained sandstones, usually buff or greenish brown in colour, with some shale beds. Ossicles of the crinoid *Balacrinus* [*Glyptocrinus*] *basalis* are locally common, and gave rise to the old name of '*Glyptocrinus* Flags'. Other fossils include the brachiopods *Reuschella horderleyensis* and *Sowerbyella soudleyensis*. Although they form a fairly distinctive lithological unit the Chatwall Flags are diachronous. In the type area they are co-extensive with the Soudleyan Stage (Dean 1958, 1960) and include beds of the same age as the upper part of the Glenburrell Beds of the Onny Valley (Fig. 31). Their upper part at Chatwall is of the same age as the Chatwall Sandstone ('Soudley Sandstone') of the Hope Bowdler area.

*Chatwall Sandstone.*   This sandstone, which usually forms a marked scarp, occurs in three distinct outcrops in which local names have been applied. In the Onny Valley area it has been called the Horderley Sandstone, and around Hope Bowdler the Soudley Sandstone. The Chatwall Sandstone in the Onny Valley (Upper and Middle Horderley Sandstone of Dean 1958) is up to 500 ft (150 m) thick and comprises massive dark green, purple and brown sandstone with some thin shelly limestone beds. Farther north, around Hope Bowdler the sandstone ('Soudley Sandstone') is thinner, about 150 ft (45 m), but of similar lithology. It is also older and lies in the upper part of the Soudleyan Stage with a non-sequence between it and the overlying *alternata* Limestone. North of Cardington Hill the Chatwall Sandstone contains some pebbly beds and beds of shale. At Chatwall itself, Dean (1960) concluded that the term 'Chatwall Sandstone' should be restricted to some 26 ft (8 m) of Lower Longvillian beds with non-sequences at top and bottom. Brachiopods are very abundant in the Chatwall Sandstone, especially *Sowerbyella soudleyensis* and species of *Dalmanella* (Fig. 23E). Trilobites include the trinucleid *Broeggerolithus nicholsoni*.

*Cheney Longville Flags.*   The Cheney Longville Flags have a broad outcrop on the dip slope of the Chatwall Sandstone scarp. The lowest beds, up to 100 ft (30 m) thick, are termed the *alternata* Limestone and consist of flags and shales interbedded with lenticular shelly limestones containing an abundance of the brachiopod *Heterorthis alternata* (Fig. 23D). The higher beds are greenish grey flags with interbedded shales and with thin shelly limestones containing many brachiopods, especially species of *Kjaerina* (Fig.

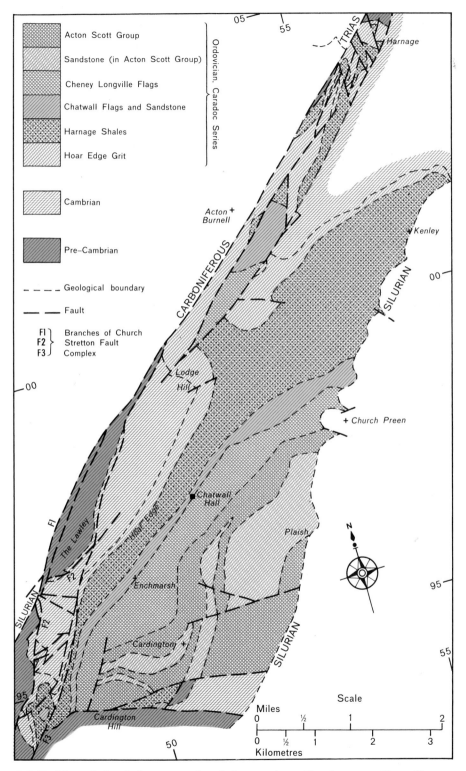

**Legend:**

Ordovician, Caradoc Series
- Acton Scott Group
- Sandstone (in Acton Scott Group)
- Cheney Longville Flags
- Chatwall Flags and Sandstone
- Harnage Shales
- Hoar Edge Grit

Cambrian

Pre–Cambrian

– – – Geological boundary

——— Fault

F1 ⎱
F2 ⎬  Branches of Church Stretton Fault Complex
F3 ⎰

Scale

Miles
0    ½    1              2

Kilometres
0   ½   1        2        3

FIG. 28. *Map of the Ordovician rocks of the northern Caradoc area, Shropshire*

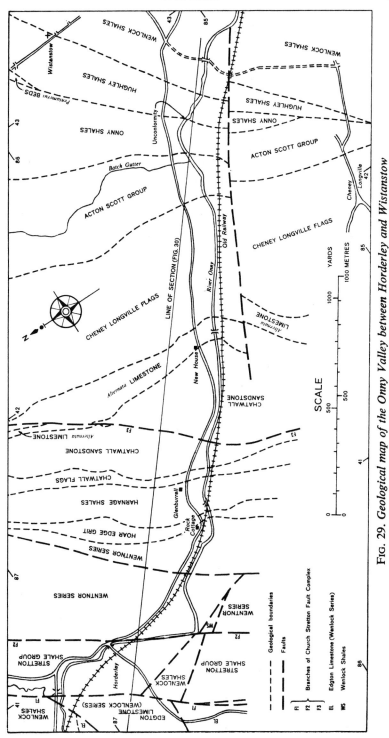

FIG. 29. *Geological map of the Onny Valley between Horderley and Wistanstow*

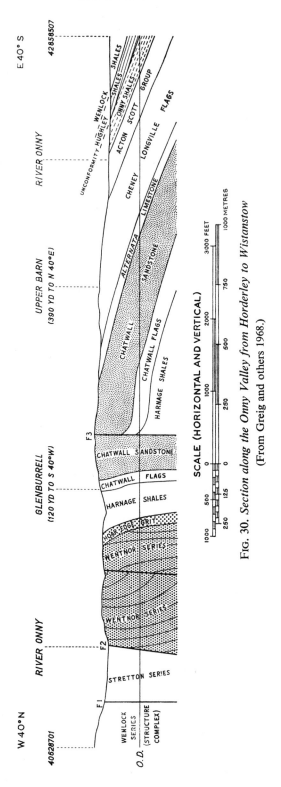

FIG. 30. *Section along the Onny Valley from Horderley to Wistanstow.*
(From Greig and others 1968.)

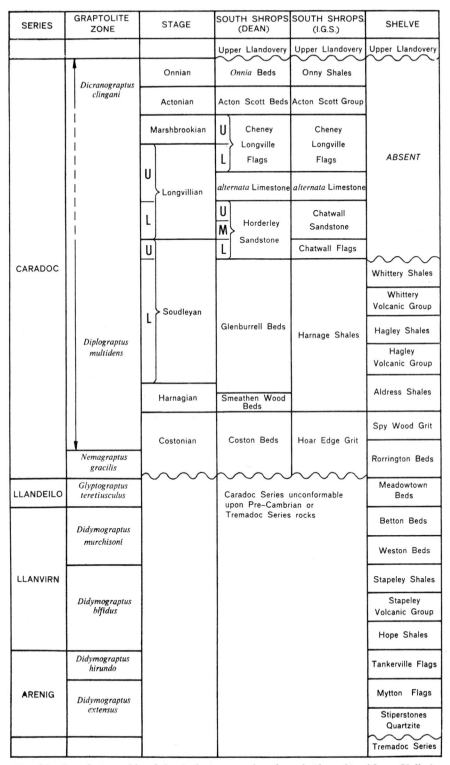

| SERIES | GRAPHOLITE ZONE | STAGE | SOUTH SHROPS. (DEAN) | SOUTH SHROPS. (I.G.S.) | SHELVE |
|---|---|---|---|---|---|
| | | | Upper Llandovery | Upper Llandovery | Upper Llandovery |
| CARADOC | *Dicranograptus clingani* | Onnian | *Onnia* Beds | Onny Shales | ABSENT |
| | | Actonian | Acton Scott Beds | Acton Scott Group | |
| | | Marshbrookian | U Cheney Longville L Flags | Cheney Longville Flags | |
| | | Longvillian U | *alternata* Limestone | *alternata* Limestone | |
| | | L | U M Horderley Sandstone | Chatwall Sandstone | |
| | | U | L | Chatwall Flags | |
| | *Diplograptus multidens* | Soudleyan L | Glenburrell Beds | Harnage Shales | Whittery Shales |
| | | | | | Whittery Volcanic Group |
| | | | | | Hagley Shales |
| | | | | | Hagley Volcanic Group |
| | | | | | Aldress Shales |
| | | Harnagian | Smeathen Wood Beds | | |
| | | Costonian | Coston Beds | Hoar Edge Grit | Spy Wood Grit |
| | *Nemagraptus gracilis* | | | | Rorrington Beds |
| LLANDEILO | *Glyptograptus teretiusculus* | | Caradoc Series unconformable upon Pre–Cambrian or Tremadoc Series rocks | | Meadowtown Beds |
| | *Didymograptus murchisoni* | | | | Betton Beds |
| | | | | | Weston Beds |
| LLANVIRN | *Didymograptus bifidus* | | | | Stapeley Shales |
| | | | | | Stapeley Volcanic Group |
| | | | | | Hope Shales |
| ARENIG | *Didymograptus hirundo* | | | | Tankerville Flags |
| | *Didymograptus extensus* | | | | Mytton Flags |
| | | | | | Stiperstones Quartzite |
| | | | | | Tremadoc Series |

FIG 31. *Correlation table of the Ordovician rocks of south Shropshire* (*Onny Valley*) *and the Shelve area*

(After Dean *in* Whittard 1955–67, fig. 10, and Greig and others 1968, table 6.)

23H). The supposed scaphopod *Tentaculites* is abundant on some bedding planes. The Flags are well exposed around Cheney Longville and Marshbrook.

*Acton Scott Group.*    North of Cardington Hill the Acton Scott Group has an extensive poorly exposed outcrop around Cardington and Plaish (Fig. 28), where it consists of some 500 ft (150 m) of yellow mudstones with two thick beds of shelly sandstone. South of Cardington Hill the lithology of the Group is rather more variable. It consists mainly of micaceous mudstones and siltstones but north of Acton Scott the highest beds are yellow flaggy sandstones and siltstones. At Acton Scott a lenticular bed of hard splintery very calcareous sandstone, the Acton Scott Limestone, is developed in the middle of the Group, forming the high ground on which the village stands. In contrast to the Cheney Longville Flags and Onny Shales, trinucleid trilobites are rare, but other trilobites include *Chasmops extensa* and *Platylichas laxatus.* The abundant brachiopod fauna includes *Cryptothyris paracyclica* and species of *Onniella* and *Reuschella.* The limits of the Group have been variously defined, but as used here it is probably approximately equivalent to Bancroft's Actonian Stage, at least in the Onny Valley (Greig and others 1968).

*Onny Shales.*    The Onny Shales have a restricted lenticular outcrop between Wistanstow and Acton Scott in the southern part of the Caradoc area. To the east they are overlain unconformably by the basal Silurian, and this unconformity can be seen in the famous section in the River Onny (Figs. 29, 30). The formation comprises grey and buff micaceous siltstones and mudstones and contains an abundant shelly fauna including various species of the trinucleid trilobite *Onnia* (Fig. 23J) and brachiopods such as *Onniella broeggeri* and '*Strophomena*' *holli.*

# References

BANCROFT, B. B. 1945. The Brachiopod Zonal indices of the Stages Costonian to Onnian in Britain. *J. Palaeont.,* **19**, 181–252.

BLYTH, F. G. H. 1938. Pyroclastic rocks from the Stapeley Volcanic Group at Knotmoor, near Minsterley, Shropshire. *Proc. Geol. Assoc.,* **49**, 392–404.

—— 1944. Intrusive rocks of the Shelve area, south Shropshire. *Quart. J. Geol. Soc. Lond.,* **99** (for 1943), 169–204.

DEAN, W. T. 1958. The Faunal Succession in the Caradoc Series of South Shropshire. *Bull. Brit. Mus. (Nat. Hist.) Geol.,* **3**, 191–231.

—— 1960. The Ordovician Rocks of the Chatwall District, Shropshire. *Geol. Mag.,* **97**, 163–71.

—— 1964. The Geology of the Ordovician and adjacent strata in the Southern Caradoc district of Shropshire. *Bull. Brit. Mus. (Nat. Hist.) Geol.,* **9**, 257–96.

—— and DINELEY, D. L. 1961. The Ordovician and Associated Pre-Cambrian Rocks of the Pontesford District, Shropshire. *Geol. Mag.,* **98**, 367–76.

GREIG, D. C., WRIGHT, J. E., HAINS, B. A. and MITCHELL, G. H. 1968. Geology of the Country around Church Stretton, Craven Arms, Wenlock Edge and Brown Clee. *Mem. Geol. Surv.*

POCOCK, R. W., WHITEHEAD, T. H., WEDD, C. B. and ROBERTSON, T. 1938. Shrewsbury District including the Hanwood Coalfield. *Mem. Geol. Surv.*

WATTS, W. W. 1885. On the Igneous and Associated Rocks of the Breidden Hills, in East Montgomeryshire and West Shropshire. *Quart. J. Geol. Soc. Lond.*, **41**, 532–46.

—— 1925. The Geology of South Shropshire. *Proc. Geol. Assoc.*, **36**, 321–63.

WEDD, C. B. 1932. Notes on the Ordovician Rocks of Bausley, Montgomeryshire. *Sum. Prog. Geol. Surv.* for 1931, **2**, 49–55.

WHITEHEAD, T. H. 1929. The Occurrence of a Rhyolite between the Ordovician and Pre-Cambrian Rocks near Habberley, Shropshire. *Sum. Prog. Geol. Surv.* for 1928, **2**, 120–5.

WHITTARD, W. F. 1931. The Geology of the Ordovician and Valentian Rocks of the Shelve Country, Shropshire. *Proc. Geol. Assoc.*, **42**, 322–39.

—— 1952. A geology of south Shropshire. *Proc. Geol. Assoc.*, **63**, 143–97.

—— 1955–67 The Ordovician Trilobites of the Shelve Inlier, West Shropshire. Parts I–IX (*Includes* Relationships of the Shelve Trilobite Faunas by W. T. Dean). *Palaeont. Soc.* [*Monogr.*].

# 5. Silurian

The Silurian System of rocks was originally explored and named in the Welsh Borderland region by Sir Roderick Murchison and, although the lower part of the great system which he named was later re-named the Ordovician System, the Welsh Borderland remains a classical one for the study of the Silurian succession and includes the type areas for the middle and upper subdivisions of the system.

The Silurian strata rest with strong unconformity on Ordovician, Cambrian and Pre-Cambrian rocks; a study of this unconformity and of facies variations and sedimentary features in the deposits themselves enables the history and palaeogeography of the Welsh Borderland during Silurian time to be clearly envisaged. A period of marine transgression across a land surface of considerable diversity in early Silurian time was followed by a long period during which a 'basin' area of fairly deep water and rapid sea-floor subsidence to the north-west of a line from Horderley to Radnor merged into a 'shelf' area to the south-east, where subsidence was slow, water much shallower and coastal shoals and small reefs were recurrent features. There were many phases, especially in the quickly subsiding 'basin' areas, when unstable masses of sediment slid upon the sea floor and came to rest with their original layering in great confusion.

The present-day scenery formed by the Silurian rocks of the 'basin' areas is mainly bold rounded hills up to 2000 ft in height, while the limestones of the 'shelf' areas give rise to conspicuous, well-wooded scarp and dip-slope features such as those of Wenlock Edge, Aymestrey and Woolhope.

The Silurian rocks fall into three main divisions:

> 3. Ludlow Series
> 2. Wenlock Series
> 1. Llandovery Series

## Llandovery Series

Llandovery rocks form narrow discontinuous outcrops against the older masses of Breidden, Shelve and the Long Mynd as well as along the northern part of the main Silurian outcrop of Wenlock Edge. They also occur in the inliers of Ledbury–Malvern, Woolhope, May Hill and Presteigne.

In Lower Llandovery times the sea covered much of Central Wales but no deposits of that period appear to have been laid down in the Welsh Borderland region which we may assume to have been a land area. With the advent of Middle and Upper Llandovery times, however, the sea gained access to the region and sandy or conglomeratic deposits were laid down along an irregular coastline around partly submerged masses of older formations. These very shallow water deposits were followed by a relatively thin sequence of more normal marine strata.

While the fossils of the Llandovery Series of the Welsh Borderland are mainly those of a shelly benthonic fauna, occasional graptolites are found in the Shropshire outcrops and these enable some of the standard graptolite zones of the Central Wales graptolitic facies to be recognized as follows:

|                    |     |     |     |                                      |
|--------------------|-----|-----|-----|--------------------------------------|
|                    | Zone of | *Monograptus griestoniensis* |     |                                      |
| UPPER LLANDOVERY   | ,, ,, |     | ,,  | *crispus*                            |
|                    | ,, ,, |     | ,,  | *turriculatus*                       |
|                    | ,, ,, |     | ,,  | *sedgwickii*                         |
| MIDDLE LLANDOVERY  | ,, ,, |     | ,,  | *convolutus*                         |

The Zone of *Monograptus crenulatus* at the top of the Llandovery has not yet been recognized.

In the main outcrop of Llandovery rocks south of the Wrekin the basal sandy deposits are termed the Kenley Grit. This is up to 150 ft (46 m) thick near Kenley; it includes conglomerates and less coarse strata and yields *Lingula* near the top. It is overlapped south of Kenley, and farther south there is successive overlap of each subdivision by that above until, south of Cheney Longville, the whole of the Llandovery Series is missing and the Wenlock Shales rest on Ordovician rocks.

After the initial Llandovery transgression there appears to have been a steady widening and deepening of the area of sedimentation throughout the remainder of Llandovery time. In the areas reached by the deeper water, fossiliferous muddy and calcareous deposits, the *Pentamerus* Beds, were laid down, the most characteristic fossil being the brachiopod *Pentamerus oblongus* (Fig. 32A). These consist of grey siltstones and silty mudstones with thin sandy layers and shelly limestone bands and they reach thicknesses up to about 400 ft (122 m) along the main outcrop. They yield many other brachiopods including *Atrypa reticularis*, *Eocoelia hemisphaerica* and *Stricklandia lens*, and corals such as *Streptelasma whittardi*, as well as graptolites indicative of the zones of *Monograptus sedgwickii* (Fig. 32C, D) and *M. turriculatus* (Fig. 32E). In Harper's Dingle south-east of the Wrekin a thin conglomerate is developed near the base of these beds composed of Uriconian pebbles mixed with quantities of the shells of *P. oblongus*, indicating the proximity of a shoreline of Uriconian rock.

On the flanks of the Long Mynd the basal Llandovery grits and conglomerates yield a fauna which suggests that they are a basal facies of the *Pentamerus* Beds. Their thickness varies from nil to more than 300 ft (91 m) and hereabouts denudation has reached a stage at which the deposits of the ancient shoreline have been uncovered at many points and, as first observed by Ramsay about 1846 and later amplified by Whittard, seem to reveal such features as the pebble beaches and sea stacks of the Llandovery sea which beat against a rugged coastline of Longmyndian rocks. Short distances away from the outcrops graptolites collected from boreholes indicate that the *Pentamerus* Beds here belong to the zones of *Monograptus convolutus* and *M. sedgwickii*. North and west of the Shelve Inlier there is a thinner development of *Pentamerus*-bearing sandy beds with a very irregular base broadly referable in horizon to the same two graptolite zones. In the Breidden Hills outcrop the approximately equivalent Cefn Beds are sandy mudstones, calcareous sandstones and occasional conglomerates, and the lower part of these beds has yielded *Stricklandia lens*.

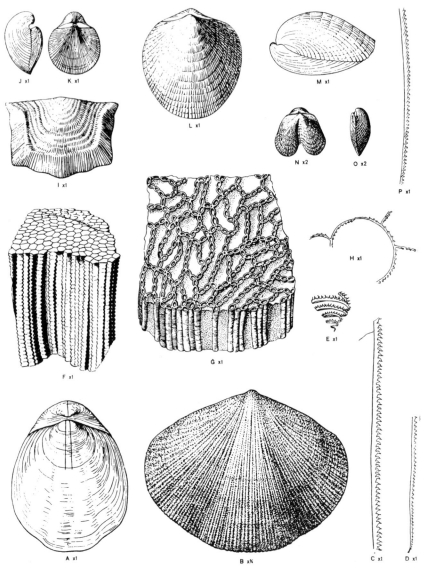

FIG. 32. *Silurian Fossils (Upper Llandovery and Wenlock Series)*
All figures natural size except B × ¾ and N, O × 2

Upper Llandovery Series: **A.** *Pentamerus oblongus* J. de C. Sowerby; **B.** *Costistricklandia lirata* (J. de C. Sowerby); **C, D.** *Monograptus sedgwickii* (Portlock); **E.** *Monograptus (Spirograptus) turriculatus* (Barrande).

Wenlock Series: **F.** *Favosites gothlandicus* Lamarck; **G.** *Halysites catenularius* (Linné); **H.** *Cyrtograptus lundgreni* Tullberg; **I.** *Leptaena* cf. *depressa* (J. de C. Sowerby); **J, K.** *Resserella* cf. *elegantula* (Dalman); **L, M.** *Atrypa reticularis* (Linné); **N, O.** *Dicoelosia biloba* (Linné); **P.** *Monograptus riccartonensis* Lapworth.

The *Pentamerus* Beds are succeeded by purple and maroon shaly mudstones, occasionally silty or sandy, with many green bands and patches. Thin calcareous laminae are common and thin beds of shelly argillaceous limestone rarely more than 2 in (50 mm) thick occur locally. These beds are known as the Hughley Shales (or Purple Shales) and they yield many brachiopods such as *Eocoelia intermedia, E. sulcata* and *Costistricklandia lirata* (Fig. 32B) as well as graptolites indicative of the zones of *Monograptus turriculatus, M. crispus* and *M. griestoniensis.* In the main outcrop from the Wrekin to Cheney Longville they are commonly between 250 ft (76 m) and 350 ft (107 m) thick.

Very similar beds occur around the Long Mynd where they vary from 150 ft (49 m) to 560 ft (184 m) in thickness, and north and west of the Shelve Inlier the same formation is represented. In the Breidden Hills outcrop the lithologically similar Buttington Shales are about 350 ft (115 m) thick and are probably the direct correlatives of the Hughley Shales.

In the outcrop of Llandovery rocks which forms the core of the Woolhope Inlier around Haugh Wood (Fig. 33), a lower group of more than 280 ft (85 m) of flaggy greyish blue calcareous sandstones and thin olive mudstones is overlain by about 20 ft (6 m) of purple and olive muddy siltstones and thin limestones, followed by 10 ft (3 m) of coarser grained olive siltstones with calcareous and crinoidal limestone bands. They contain brachiopods such as *Cyrtia exporrecta, Leangella segmentum* and the large *Costistricklandia lirata* as well as corals such as *Goniophyllum pyramidale* and *Chonophyllum? [Ptychophyllum] patellatum.* In the top 10 ft (3 m) there is a thin band of limestone crowded with the remains of the peculiar crinoid *Petalocrinus.*

The Llandovery rocks which form May Hill and Huntley Hill in the May Hill Inlier consist of more than 600 ft (183 m) of thickly bedded coarse grey or pink sandstones and conglomerates with *Eocoelia hemisphaerica,* which pass up into some 500 ft (152 m) of flaggy yellowish sandstones and siltstones with subordinate shales. The upper group is very fossiliferous there being abundant *Costistricklandia lirata* associated with *Stricklandia lens, Pentamerus oblongus, Leptostrophia compressa,* corals and trilobites. The highest beds are finer grained and calcareous and they include the *Petalocrinus* Limestone as at Woolhope.

In the Ledbury–Malvern Inlier Llandovery rocks crop out extensively in the Eastnor Park area and in a narrow belt from north of Herefordshire Beacon to Old Storridge Common. There is also a small inlier at Ankerdine Hill. The basal Llandovery beds are brown and red siltstones and sandstones locally with *Lingula.* These are overlain by more than 200 ft (65 m) of coarse pink sandstones and grey and purple conglomerates the latter well developed at Cowleigh Park. The conglomerate pebbles are mostly of white quartz and of Cambrian and older rocks showing that a landmass composed of Malvernian and Cambrian rocks was being eroded. Grey and green siltstones with brown shales succeed the conglomerates and a rich Upper Llandovery fauna with *Costistricklandia lirata* is usually present either just below or just above the lithological change. The highest beds are green and purple shales with thin bands of limestone, among which the *Petalocrinus* Limestone can be traced over most of the area of the inlier. These upper formations are from 250 to 900 ft (76–274 m) thick and the higher measures, which pass up into the Woolhope Limestone, have been variously named the Woolhope Shale, *Stricklandia* Beds or Tarannon Shales.

The small sharply anticlinal inlier of Llandovery rocks immediately south of Presteigne consists of pebbly sandstones, probably more than 70 ft (21 m) thick, which have yielded *Pentamerus oblongus, Eocoelia hemispherica* and *Stricklandia lens.* A few miles to the south-west, at Old Radnor, no strata of Llandovery age occur between the base of the Dolyhir (Woolhope) Limestone and the underlying Longmyndian.

## Wenlock Series

The 'shelf' facies of the Wenlock Series is well developed all along the main Silurian outcrop from Much Wenlock to Aymestrey as well as in the Silurian inliers of Ledbury–Malvern, Woolhope, May Hill and Usk. There are also small outcrops in the Abberley Hills area east of Rodge Hill and in Abberley Park. The classification of these beds is based on the developments at Wenlock Edge and Woolhope (Fig. 33), and three main divisions are recognized as follows:

> Wenlock Limestone
> Wenlock Shale
> Woolhope Limestone

The fauna of the Woolhope Limestone is mainly restricted to brachiopods, and a few corals and trilobites, but that of the Wenlock Shale and Wenlock Limestone is much richer. As well as numerous brachiopods, trilobites and corals, graptolites are not uncommon in the Wenlock Shale and their presence facilitates correlation with the 'basin' facies of other areas.

The 'basin' facies of the Wenlock Series, which is present to the west and north-west of the Horderley to Radnor line, is represented by grey shales and calcareous siltstones with a fairly abundant fauna of both graptolites and shelly fossils. The sequence is best known in the Long Mountain where more than 1500 ft (457 m) of measures are divisible into four graptolite zones as follows:

> Zone of *Cyrtograptus lundgreni*
>    ,,   ,,       ,,    *ellesae*
>    ,,   ,,       ,,    *linnarssoni*
>    ,,   ,,  *Monograptus riccartonensis*

Elsewhere zones of *Cyrtograptus centrifugus* and *Cyrtograptus murchisoni* are recognized below the Zone of *M. riccartonensis* and a Zone of *Cyrtograptus rigidus* may be present between the zones of *M. riccartonensis* and *C. linnarssoni.* Also it has recently been advocated that above the *C. lundgreni* Zone, the Zone of *Monograptus vulgaris* (*M. ludensis*), hitherto regarded as represented in the 'shelf' sequence by the basal part of the Lower Ludlow Shales (Elton Beds), should be allocated to the Wenlock Series. It now seems that the whole of the Wenlock Limestone and nearly 200 ft (61 m) of the underlying Wenlock Shale probably belong to the zone, therefore its inclusion within the Wenlock would seem to be appropriate. In the Long Mountain succession shelly fossils such as '*Chonetes*' *minimus, Resserella* cf. *elegantula* (Fig. 32J, K), *Dicoelosia biloba* (Fig. 32N, O) and *Dalmanites caudatus* are common in calcareous horizons in strata referred to the zones of *C. linnar-*

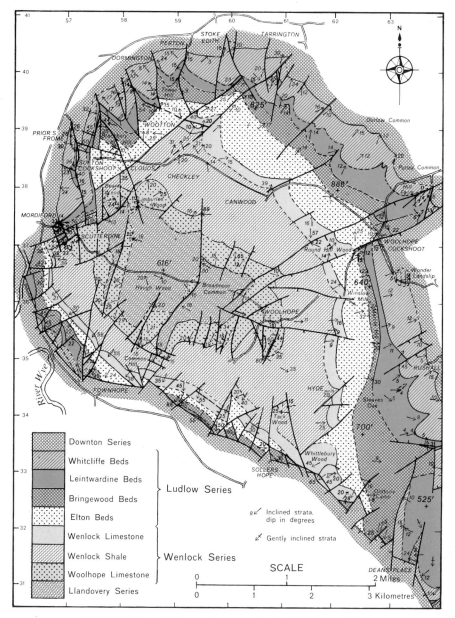

FIG. 33. *Map of the main part of the Woolhope Inlier*
*(After Squirrell and Tucker)*

*ssoni, C. lundgreni* and *M. vulgaris.* Argillaceous limestones, the possible correlatives of those found in the 'shelf' areas are known only in the vicinity of Edgton, immediately west of the Church Stretton Fault. Presumably the deeper waters of the 'basin' areas were less favourable to the limestone-building fauna of the shallower waters to the south-east.

*Woolhope Limestone.* In the inliers of Woolhope, Ledbury–Malvern and May Hill the Woolhope Limestone varies from about 50 to 200 ft (15–61 m) in thickness. At Woolhope it is mainly grey, nodular, argillaceous limestone or thickly bedded limestone, but at Ledbury–Malvern and May Hill two or more horizons of nodular argillaceous limestone may be separated by as much as 150 ft (46 m) of olive brown calcareous shale. The fauna is poor, the commonest fossils being brachiopods and corals. Among the former occur *Resserella* cf. *elegantula, Atrypa reticularis* (Fig. 32L, M), *Leptaena depressa* (Fig. 32I) and *Plectodonta transversalis* and the commonest corals are *Favosites gothlandicus* (Fig. 32F) and *Halysites catenularius* (Fig. 32G). The trilobite *Bumastus barriensis* occurs at Woolhope and Malvern.

In the northern areas of Wenlock Edge and Buildwas in the Severn Valley the position of the Woolhope Limestone is occupied by the shales of the Buildwas Beds, which succeed the Hughley Shales of the Llandovery Series. In the west, at Old Radnor and Nash Scar near Presteigne a massive development of algal (*Solenopora*) limestone (the Dolyhir Limestone) is generally considered to be a reef facies of the Woolhope Limestone. Its abundant fauna, however, shows affinities with that of the Wenlock Shale, which it may in part represent. At Old Radnor (Fig. 34) a basal conglomerate is present resting on and containing rolled fragments of Longmyndian rocks. Near Presteigne the limestone rests on pebbly sandstones of the Llandovery Series.

*Wenlock Shale.* The thickness of the Wenlock Shale is remarkably constant throughout most of the region. It averages about 1000 ft (305 m) along the main outcrop and at Woolhope, but it thins somewhat to 700 to 800 ft (213–244 m) at Usk, May Hill and in the Ledbury–Malvern Inlier. The formation consists of soft olive-grey calcareous mudstones and fine calcareous siltstones with layers of calcareous nodules most commonly at the base and towards the top. The last mentioned are most fully developed in the top 100 ft (30 m) or so in the Wenlock area where they are separated under the local name Tickwood Beds. Although seldom well exposed in the usually low lying terrain of its outcrop, good exposures of the Wenlock Shale are usually fairly fossiliferous, with small brachiopods predominant. The latter include *Atrypa reticularis, Resserella* cf. *elegantula, Plectodonta transversalis, Dicoelosia biloba, Leptaena depressa, Howellella elegans* etc. Small turbinate corals are fairly common and there are usually many fragments of the trilobite *Dalmanites.* A number of exposures along the main outcrop have yielded *Monograptus flemingii, M. priodon* and *M. vomerinus;* and boreholes into the outcrops south of the Long Mynd have yielded graptolites of the *M. riccartonensis* (Fig. 32P) and *C. rigidus* zones.

*Wenlock Limestone.* Along Wenlock Edge the Wenlock Limestone averages about 100 ft (30 m) in thickness, but south-westwards it thickens to as much

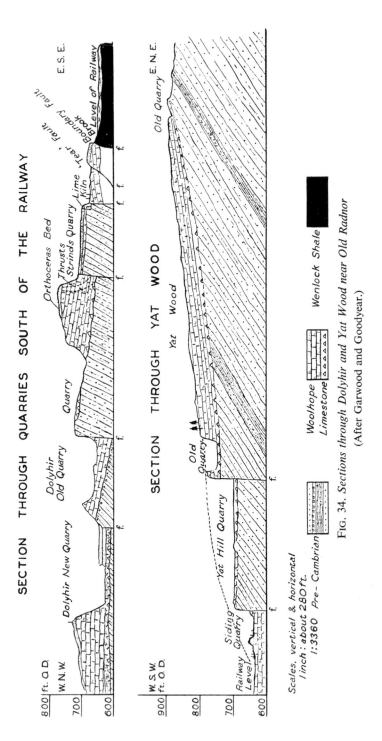

SECTION THROUGH QUARRIES SOUTH OF THE RAILWAY

SECTION THROUGH YAT WOOD

Scales, vertical & horizontal
1 inch: about 280 ft.
1:3360

Woolhope Limestone   Wenlock Shale

Pre-Cambrian

FIG. 34. *Sections through Dolyhir and Yat Wood near Old Radnor*
(After Garwood and Goodyear.)

**Plate V**

A. Wrekin Quartzite,
Ercall Quarry,
Wellington, Shropshire
(A 11102)

(*For full explanation see p. x*)

B. Pontesford Hill,
Shropshire, viewed
from the south
(A 11103)

# Plate VI

A. Reef development in
   Wenlock Limestone,
   Lilleshall Quarry
   near Presthope
   (A 9543)

*(For full explanation see p. x)*

B. Quarry face in
   bedded Wenlock
   Limestone near
   Moorwood, Craven
   Arms
   (A 9540)

as 450 ft (137 m) of interbedded calcareous and argillaceous sediments at Leintwardine. At Woolhope it is about 150 ft (46 m) thick and at May Hill and in the Ledbury–Malvern and Abberley inliers it shows similar extremes of variation to those between Wenlock Edge and Leintwardine. At Usk it is scarcely more than 40 ft (12 m) thick.

The bulk of the limestone is grey and nodular, varying from thinly bedded and argillaceous to thickly bedded and highly calcareous rock which has been locally much quarried (Plate VIB). In these more limy facies, beds of shell fragmental limestone are very common. Calcareous shale and mudstone occur as partings throughout the major and minor limestones as well as forming beds 50 ft (15 m) or more thick which separate the main limestone horizons. Bands of pisolitic limestone, usually less than a foot thick, are fairly common at May Hill and in the Malverns.

An important feature of the Wenlock Limestone is the occurrence of 'ballstone' reefs (bioherms) apparently in any part of the limestone though the lower part at Wenlock Edge is notable for their abundance (Plate VIA). They are usually flat-bottomed lenticular masses of unstratified limestone up to 80 ft (24 m) thick and composed of large and irregular colonies mainly of corals and stromatoporoids in their position of growth set in a matrix of grey and green calcite-mudstone. The framework of the reefs was probably built by upgrowth, from favourable loci on the sea bottom, of corals and stromatoporoids, and the matrix may have formed as a chemical precipitate from the decomposition of algal material which evidently abounded in the colony. The surrounding stratified limestone contains a rich fauna of detrital coral material suggesting that the reefs were subject to, and probably partly destroyed by wave action before their final burial.

Some common corals in the 'ballstone' reefs are *Heliolites interstinctus*, *H. parvistellus* and *H. parvistellus* var. *caespitosus*; other corals, common throughout the Wenlock Limestone, include *Favosites gothlandicus* and *Halysites catenularius*. Among brachiopods in the Wenlock Limestone, *Atrypa reticularis*, *Leptaena* cf. *depressa*, *Resserella* cf. *elegantula*, *Dalejina hybrida*, *Strophonella euglypha* and *Dicoelosia biloba* are usually abundant and many other species are very common; there are usually innumerable crinoid fragments and many bryozoa but mollusca are relatively few. Trilobites such as *Calymene spp.* and *Dalmanites spp.* are locally common.

## Ludlow Series

The rocks of the Ludlow Series can be even more readily designated as of 'shelf' or of 'basin' facies than those of the Wenlock, and they exhibit striking increases in thickness from south-east to north-west in the vicinity of the Church Stretton Fault (Fig. 35). The traditional tripartite division of the 'shelf' facies of the Ludlow Series into Lower Ludlow, Aymestry Limestone and Upper Ludlow is now less useful for correlation than the more recent fourfold classification based on detailed studies of the Ludlow succession of the Ludlow Anticline (Plates I and VII). A simplified version of the latest classification, which has now proved to be applicable to most of the 'shelf' facies outcrops, and its relationship to the older classification is as follows:

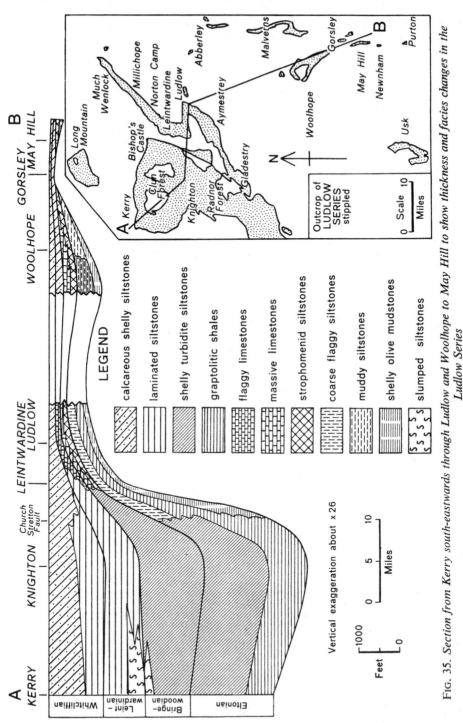

FIG. 35. *Section from Kerry south-eastwards through Ludlow and Woolhope to May Hill to show thickness and facies changes in the Ludlow Series*

(Adapted from Holland and Lawson.)

| | |
|---|---|
| Whitcliffe Beds | ⎫ Upper Ludlow Flags |
| Leintwardine Beds | ⎬ |
| Bringewood Beds | Aymestry Limestone |
| Elton Beds | Lower Ludlow Shales |

In the graptolite-bearing strata below the Whitcliffe Beds five zones were formerly recognized; these are in descending order, the zones of *Monograptus (Saetograptus) leintwardinensis, M. tumescens, M. scanicus, M. nilssoni,* and *M. vulgaris.* The strata comprising the *M. vulgaris* Zone at the base (see p. 62) locally pass laterally into the top part of the Wenlock Limestone and it has been suggested that this Zone should be referred to the Wenlock. Furthermore, the *M. tumescens* Zone is of doubtful validity while the *M. nilssoni* and *M. scanicus* zones are in many places difficult to separate. Recent workers therefore tend to recognize within the Ludlow rocks of the Welsh Borderland a *nilssoni-scanicus* Zone followed by strata with graptolites such as *M. salweyi, M. varians* and *M. leintwardinensis incipiens* overlain by a well defined Zone of *M. leintwardinensis.*

### 'Shelf' Facies

*Elton Beds.* In the Ludlow area the Lower Elton Beds are pale olive calcareous silty mudstones with a shelly fauna which includes *Resserella* cf. *elegantula* and *Dicoelosia biloba*; the Middle Elton Beds are olive-grey shaly and flaggy siltstones with graptolites of the *M. nilssoni* and *M. scanicus* zones, and the Upper Elton Beds are harder calcareous siltstones with '*Chonetes lepisma*' and *Monograptus (Pristiograptus) tumescens.* The thickness of the Elton Beds varies from less than 400 ft (122 m) below Wenlock Edge to more than 800 ft (244 m) around Leintwardine; in the Usk Inlier they are 500 to 700 ft (152–213 m) thick and at Woolhope some 600 ft (183 m) in the north-west, thinning to 100 ft (30 m) or so in the south-east. They are thin at May Hill and absent at Gorsley. In the Ledbury–Malvern Inlier they vary from about 200 ft (61 m) to over 600 ft (183 m) and they are just over 300 ft (91 m) thick at Abberley.

*Bringewood Beds.* The Lower Bringewood Beds at Ludlow are mainly calcareous siltstones and silty limestones with abundant large brachiopods particularly strophomenids. The Upper Bringewood Beds are irregularly bedded flaggy and silty limestones with fairly common shell bands rich in remains of the large brachiopod *Conchidium knightii* (Fig. 36D and Plate VIIIA); they also contain numerous other brachiopods and many compound corals. The thickness of the Bringewood Beds is about 200 ft (61 m) along Wenlock Edge but it varies from maxima of about 250 ft (76 m) at Ludlow and Usk and 280 ft (85 m) at Woolhope to 50 ft (15 m) at May Hill, and to nil at Gorsley and at places near Leintwardine where submarine channelling during Lower Leintwardine Beds times seems to have removed the lime sediment. In the Ledbury–Malvern Inlier they vary from about 50 to 150 ft (15–46 m) and they are just over 100 ft (30 m) thick at Abberley.

*Leintwardine Beds.* The Lower Leintwardine Beds of the Ludlow Anticline are olive-grey calcareous flaggy and shaly siltstones with layers of thin shelly limestone. Brachiopods such as *Dayia navicula* (Fig. 36E, F), *Sphaerirhynchia wilsoni* (Fig. 36B, C) and *Isorthis orbicularis* are abundant and

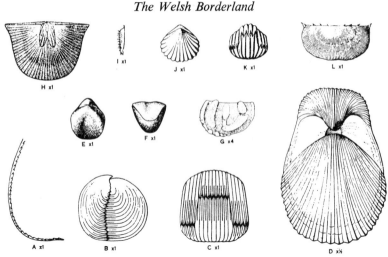

FIG. 36. *Silurian Fossils (Ludlow Series)*
All figures natural size except D × ½ and G × 4

Elton Beds: **A.** *Monograptus (Pristiograptus) nilssoni* (Barrande). Bringewood Beds: **B, C.** *Sphaerirhynchia wilsoni* (J. Sowerby); **D.** *Conchidium knightii* (J. Sowerby). Leintwardine Beds: **E, F.** *Dayia navicula* (J. de C. Sowerby); **G.** *Neobeyrichia lauensis* (Kiesow); **H.** *Shaleria ornatella* (Davidson); **I.** *Monograptus (Saetograptus) leintwardinensis* Lapworth. Whitcliffe Beds: **J, K.** '*Camarotoechia*' *nucula* (J. de C. Sowerby); **L.** *Protochonetes ludloviensis* Muir-Wood.

*Lingula lata* is common; the zonal graptolite *M. (Saetograptus) leintwardinensis* (Fig. 36I) is also fairly common. The Upper Leintwardine Beds are in places of similar lithology but elsewhere they include more thickly bedded flaggy siltstones forming a transition to the Whitcliffe Beds. The brachiopods *Shaleria ornatella* (Fig. 36H) and *Chonetoidea grayi* are usually abundant and the large ostracod *Neobeyrichia lauensis* (Fig. 36G) is very characteristic of this horizon. *M. (S.) leintwardinensis* also occurs. The thickness of the Leintwardine Beds at Wenlock Edge, Ludlow, Usk and Woolhope is mainly between 100 and 200 ft (30–61 m) but the beds thin to less than 50 ft (15 m) at May Hill and to only a few feet at Gorsley. They are 50 to 100 ft (15–30 m) thick in the Ledbury–Malvern Inlier and about 80 ft (24 m) at Abberley.

*Whitcliffe Beds.* The Lower Whitcliffe Beds at Ludlow are thickly and irregularly bedded, grey, calcareous, coarse to medium grained siltstones. The long-ranging brachiopod *Dayia navicula* occurs only in the lower part and the chief characteristic fossils are *Protochonetes ludloviensis* (Fig. 36L), *Salopina lunata*, '*Camarotoechia*' *nucula* (Fig. 36J, K) and *Beyrichia torosa*. They pass up into well-bedded, olive-grey, micaceous, medium to coarse siltstones, the Upper Whitcliffe Beds, which yield a similar but more abundant fauna especially rich in *Salopina lunata*. The highest beds contain the brachiopod *Howellella elegans* and there are bands of shelly limestone rich in gastropods. The Whitcliffe Beds are mainly less than 200 ft (61 m) in thickness at Usk, Woolhope, Ludlow and Wenlock Edge but they thicken considerably to the west of Ludlow, and at Leintwardine they are 500 to 600 ft (152–183 m) thick. They are less than 50 ft (15 m) thick at May Hill and even more attenuated at Gorsley. In the Ledbury–Malvern Inlier they are 50 to 100 ft (15–30 m) thick and about 100 ft (30 m) thick at Abberley.

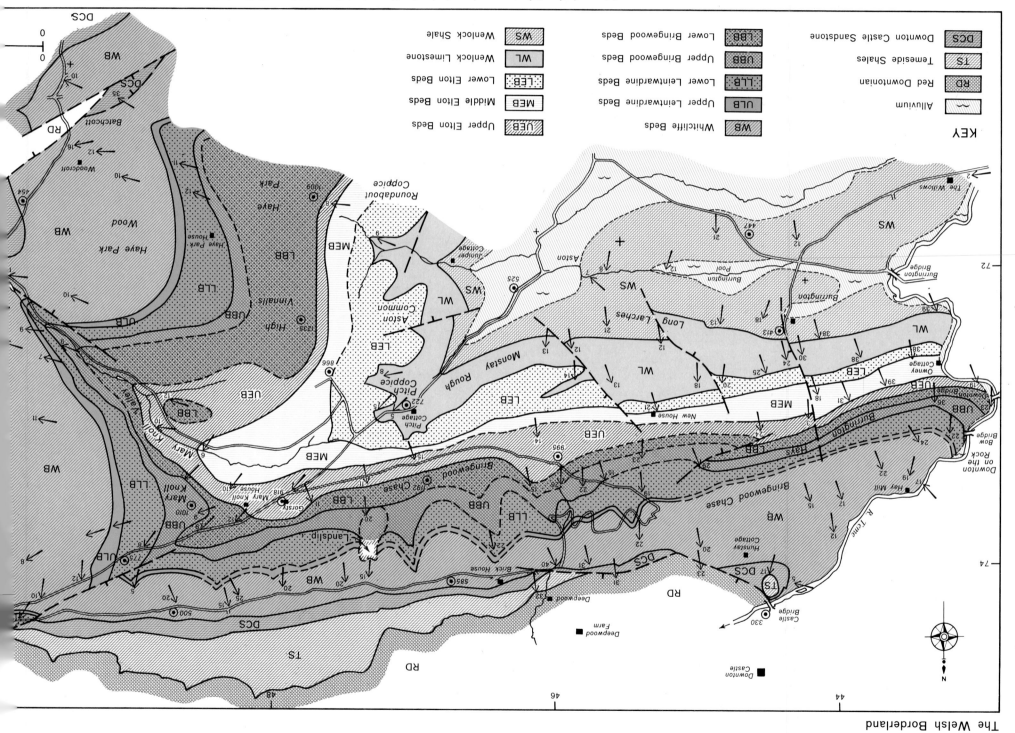

*Map of the Silurian rocks of the Ludlow Anticline (After Holland, Lawson and Walmsley)*

### 'Basin' Facies

The thick 'basin' facies of the Ludlow Series is fully represented in the extensive mountainous area around the Downtonian outliers of Clun Forest, as well as constituting most of the Long Mountain to the north and Radnor Forest to the south. It is difficult to generalize the stratigraphy of such a thick and complex formation but the broad features of the succession in these areas may be summarized as follows:

|  |  | *Generalized thickness* |
|---|---|---|
|  |  | *feet*    (*metres*) |
| 6. | Mainly siltstones with a Whitcliffe Beds fauna . | 500 to 1000 (152–305) |
| 5. | Siltstones and mudstones with few fossils other than *Dayia navicula* and *Cardiola interrupta* . | 1000 to 1500 (305–457) |
| 4. | Shales and mudstones with *Lingula lata*, *Chonetoidea grayi*, *Neobeyrichia lauensis* and *Monograptus* (*Saetograptus*) *leintwardinensis* . . | 500 to 750 (152–229) |
| 3. | Siltstones and mudstones with *Sphaerirhynchia wilsoni*, *Shaleria ornatella*, *Actinopterella tenuistriata* and graptolites of the *M. chimaera* and *M.* (*S.*) *leintwardinensis* var. *incipiens* types . | 1500 to 2000 (457–610) |
| 2. | Siltstones and mudstones with '*Chonetes lepisma*', *Isorthis orbicularis* and graptolites mainly referable to the *M. scanicus* Zone . . . | 1000    (305) |
| 1. | Shales with graptolites mainly referable to the *M. nilssoni* Zone, and small brachiopods such as *Chonetoidea grayi* . . . . . | 500    (152) |

In many places in the Clun Forest and Radnor Forest areas most of groups 3 and 4 in the above sequence are replaced by varying thicknesses, up to about 2500 ft (762 m) of contemporaneously deformed strata (Plate

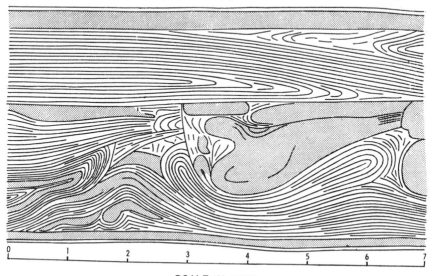

SCALE IN FEET

FIG. 37. *Sketch of part of a slumped sheet in 'basin' facies siltstones of the Ludlow Series*

VIIIв and Fig. 37) which record innumerable occasions during sedimentation when extensive areas of unstable silt and mud slipped and rolled upon the sea bed. These movements may be part of the same sedimentological disturbances as those which, at Leintwardine on the 'shelf' margin, caused channelling of the Bringewood and Elton Beds during Lower Leintwardine Beds times.

# References

ALEXANDER, FRANCES E. S. 1936. The Aymestry Limestone of the Main Outcrop. *Quart. J. Geol. Soc. Lond.*, **92**, 103–115.

CANTRILL, T. C. 1917. On a boring for coal at Presteign, Radnorshire. *Geol. Mag.*, **54**, 481–92.

DAS GUPTA, T. 1932. The Salopian graptolite shales of the Long Mountain and similar rocks of Wenlock Edge. *Proc. Geol. Assoc.*, **43**, 325–63.

DAVIS, J. E. 1850. On the age and position of the limestone of Nash near Presteign, South Wales. *Quart. J. Geol. Soc. Lond.*, **6**, 432–437.

EARP, J. R. 1938. The higher Silurian rocks of the Kerry district, Montgomeryshire. *Quart. J. Geol. Soc. Lond.*, **94**, 125–60.

—— 1940. The geology of the South-Western part of Clun Forest. *Quart. J. Geol. Soc. Lond.*, **96**, 1–11.

ELLES, GERTRUDE L. 1900. Zonal classification of the Wenlock shales of the Welsh Borderland. *Quart. J. Geol. Soc. Lond.*, **56**, 370–414.

—— and SLATER, IDA L. 1906. The Highest Silurian rocks of the Ludlow District. *Quart. J. Geol. Soc. Lond.*, **62**, 195–221.

GARWOOD, E. J. and GOODYEAR, EDITH. 1919. On the geology of the Old Radnor District with special reference to an Algal Development in the Wenlock Limestone. *Quart. J. Geol. Soc. Lond.*, **74**, 1–30.

GREIG, D. C., WRIGHT, J. E., HAINS, B.A. and MITCHELL, G. H. 1968. Geology of the country around Church Stretton, Craven Arms, Wenlock Edge and Brown Clee. *Mem. Geol. Surv.*

GROOM, T. T. 1899. The geological structure of the Southern Malverns and of the adjacent district to the west. *Quart. J. Geol. Soc. Lond.*, **55**, 129–69.

—— 1900. On the geological structure of portions of the Malvern and Abberley Hills. *Quart. J. Geol. Soc. Lond.*, **56**, 138–97.

HOLLAND, C. H. 1959. The Ludlovian and Downtonian rocks of the Knighton District, Radnorshire. *Quart. J. Geol. Soc. Lond.*, **114**, 449–82.

—— LAWSON, J. D. and WALMSLEY, V. G. 1963. The Silurian Rocks of the Ludlow District, Shropshire. *Bull. Brit. Mus. (Nat. Hist.) Geol.*, **8**, 93–171.

—— RICKARDS, R. B. and WARREN, P. T. 1969. The Wenlock graptolites of the Ludlow district, Shropshire, and their stratigraphical significance. *Palaeontology*, **12**, 663–83.

LAPWORTH, C. 1879. On the tripartite classification of the Lower Palaeozoic rocks. *Geol. Mag.*, (2), **9**, 563–6.

LAWSON, J. D. 1954. The Silurian succession at Gorsley (Herefordshire). *Geol. Mag.*, **91**, 227–37.

—— 1955. The geology of the May Hill Inlier. *Quart. J. Geol. Soc. Lond.*, **111**, 85–116.

A. Shell-bed of *Conchidium knightii*, in the Bringewood Beds, View Edge Quarry, Craven Arms

(A 9535)

(*For full explanation see p. x*)

**Plate VIII**

B. Recumbent fold in slumped 'basin' facies siltstones of the Ludlow Series

(A 11104)

A. The scenery of the Abberley Hills; view looking north-east from Hamcastle Farm

(A 7224)

(*For full explanation see pp. x–xi*)

**Plate IX**

B. Ludlow Castle and the River Teme

(A 6224)

MITCHELL, G. H., POCOCK, R. W. and TAYLOR, J. H. 1962. Geology of the country around Droitwich, Abberley and Kidderminster. *Mem. Geol. Surv.*

MURCHISON, R. I. 1834. On the structure and classification of the Transition Rocks of Shropshire, Herefordshire and part of Wales etc. *Proc. Geol. Soc. Lond.,* **2**, 13–18.

—— 1835. On the Silurian System of Rocks. *Lond. and Edinb. Phil. Mag.,* **7**, 46–52.

—— 1839. The *Silurian System*. London.

PHIPPS, C. B. and REEVE, F. A. E. 1967. Stratigraphy and geological history of the Malvern, Abberley and Ledbury hills. *Geol. J.,* **5**, 339–68.

POCOCK, R. W., WHITEHEAD, T. H., WEDD, C. B. and ROBERTSON, T. 1938. Shrewsbury District including the Hanwood Coalfield. *Mem. Geol. Surv.*

ROBERTSON, T. 1927. The Highest Silurian Rocks of the Wenlock District. *Sum. Prog. Geol. Surv.* (for 1926), 80–97.

SHERGOLD, S. H. and SHIRLEY, J. 1968. The faunal stratigraphy of the Ludlovian rocks between Craven Arms and Bourton, near Much Wenlock, Shropshire. *Geol. J.,* **6**, 119–138.

SQUIRRELL, H. C. and TUCKER, E. V. 1960. The geology of the Woolhope inlier, Herefordshire. *Quart. J. Geol. Soc. Lond.,* **116**, 139–85.

—— and DOWNING, R. A. 1969. Geology of the South Wales Coalfield, Pt. 1. The Country around Newport (Mon.). 3rd edit. *Mem. Geol. Surv.*

STAMP, L. D. 1919. The highest Silurian rocks of the Clun Forest District (Shropshire). *Quart. J. Geol. Soc. Lond.,* **74**, 221–46.

STRAW, S. H. 1937. The higher Ludlovian rocks of the Builth District. *Quart. J. Geol. Soc. Lond.,* **93**, 406–53.

WALMSLEY, V. G. 1959. The geology of the Usk inlier (Monmouthshire). *Quart. J. Geol. Soc. Lond.,* **114**, 483–521.

WELCH, F. B. A. and TROTTER, F. M. 1961. Geology of the country around Monmouth and Chepstow. *Mem. Geol. Surv.*

WHITAKER, J. H. McD. 1962. The geology of the area around Leintwardine. *Quart. J. Geol. Soc. Lond.,* **118**, 319–51.

WHITTARD, W. F. 1928. The stratigraphy of the Valentian rocks of Shropshire. The Main Outcrop. *Quart. J. Geol. Soc. Lond.,* **83**, 737–59.

—— 1932. The stratigraphy of the Valentian rocks of Shropshire. The Longmynd—Shelve and Breidden Outcrops. *Quart. J. Geol. Soc. Lond.,* **88**, 859–902.

WOOD, ETHEL M. R. 1900. The Lower Ludlow formation and its graptolite fauna. *Quart. J. Geol. Soc. Lond.,* **56**, 415–92.

ZIEGLER, A. M., COCKS, L. R. M. and McKERROW, W. S. 1968. The Llandovery Transgression of the Welsh Borderland. *Palaeontology,* **11**, 736–82.

# 6. Old Red Sandstone

Stratigraphical continuity between the rocks of the so-called 'Old Red System' and those of the underlying 'Transition' or 'Grauwacke' series was first demonstrated in the Welsh Borderland by Sir R. Murchison. Thus, as with the Silurian, the region is a classical one for the study of the Old Red Sandstone sequence and it includes the type areas for much of the succession.

The changes in sedimentation which marked the transition from Silurian to Old Red Sandstone facies suggest that, from early Old Red Sandstone times earth movements were disturbing areas north of the Welsh Borderland region. There is, however, little evidence for the existence of land in Central and North Wales during much of Lower Old Red Sandstone times because the marine and brackish-water Downtonian sediments of Montgomeryshire are thicker and more fully developed than their counterparts along the main Shropshire–Herefordshire outcrop.

During Lower Old Red Sandstone times the sea occupying the Welsh Borderland region was gradually replaced by estuarine and fresh water in which a great thickness of sediment accumulated until the process was ended by the onset of the final phase of the Caledonian earth movements. The area was uplifted and extensively eroded during Middle Old Red Sandstone times, then there was renewed subsidence and the Upper Old Red Sandstone was deposited unconformably upon the erosion surface.

Fossils of the Lower Old Red Sandstone include the very interesting early vascular plants which pioneered the colonization of the land by holophytic organisms. They also include the remarkable armoured jawless fish, the ostracoderms, which were among the earliest vertebrates to adapt themselves to a freshwater environment.

## Lower Old Red Sandstone

The Lower Old Red Sandstone occupies very large areas of Shropshire, Herefordshire, Brecknockshire and Monmouthshire as well as small areas of adjacent counties. It reaches a maximum thickness of about 6000 ft (1829 m) and it is divided into three subdivisions or series as follows:

Brecon Series
Ditton Series
Downton Series

The succession of strata comprised within these subdivisions in various parts of the Welsh Borderland is shown on Fig. 38.

### Downton Series

Apart from the basal strata which flank the Silurian hills the rocks of the Downton Series occupy most of the gentler terrain underlain by the Lower Old Red Sandstone, such as Corve Dale, the Middle Teme Valley and the lowlands and plains of Hereford. Large outliers make elevated tracts

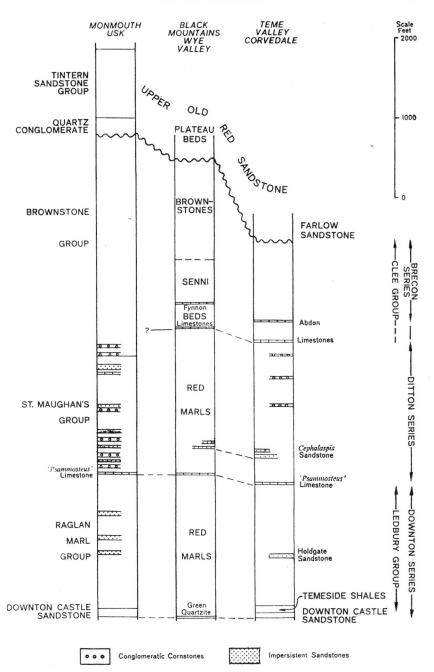

FIG. 38. *Generalized sections of the Old Red Sandstone*

in Clun Forest and there is a small outlier of the lowest 150 ft (46 m) of the Downton Series on the Long Mountain. The rocks of the Downton Series are divided into a lower, mainly greenish or yellowish grey group, the Temeside Group and an upper, mainly red group, the Ledbury Group. The Temeside Group includes, at the base, the well-known Ludlow Bone Bed, overlain by a foot or two of fossiliferous shaly strata which are represented west of the Church Stretton Fault by 20 to 40 ft (6–12 m) of very fossiliferous silty strata known as the *Platyschisma helicites* Beds (Fig. 39). Above the basal strata lies the well-known Downton Castle Sandstone followed in the north by the Temeside Shales.

*Ludlow Bone Bed.* This is a brown sandy or silty bed packed with organic fragments cemented by a small amount of calcite. It varies in thickness from an inch or less to a foot or more in various parts of the region, and is locally two or more bands separated by a few inches of mudstone. The organic fragments are mainly dermal plates, spines and teeth of ostracoderms and fragments of brachiopods, crustacea and eurypterids, and they are to some

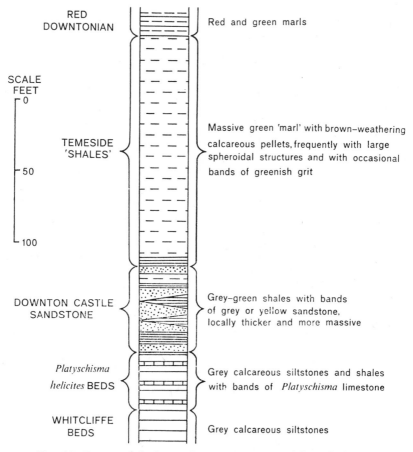

FIG. 39. *Section of the Lower Downtonian strata of the Felindre Basin*

extent rolled and worn. Well-preserved valves of characteristic Whitcliffe Beds brachiopods also occur in the bone bed.

*Platyschisma helicites Beds.* These are grey flaggy and shaly siltstones with 'posts' of tough calcareous siltstone and layers of highly micaceous ripple-bedded siltstone. Most characteristic of the group are bands up to several

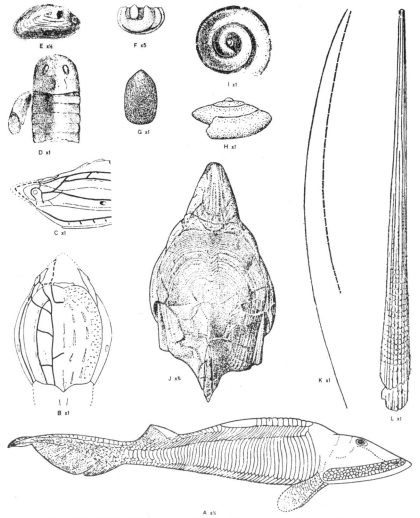

FIG. 40. *Old Red Sandstone Fossils* (*Downton and Ditton Series*)
All figures natural size except A, E × ½, J × ¾ and F × 5

Downton Series: **A.** Restoration of *Hemicyclaspis murchisoni* (Egerton) after Stensiö; **B, C.** Restorations of carapace of *Traquairaspis pococki* (White) after White; **D.** *Hughmilleria* [*Eurypterus*] *pygmaea* (Salter); **E.** *Modiolopsis complanata* (J. de C. Sowerby); **F.** *Frostiella sp.* [*Kloedenia wilkensiana* of authors]; **G.** *Lingula cornea* J. de C. Sowerby; **H, I.** '*Platyschisma*' *helicites* (J. de C. Sowerby).

Ditton Series: **J.** Dorsal shield of *Pteraspis rostrata* (Agassiz) *trimpleyensis* White; **K, L.** Dorsal fin-spine of *Nodonchus bambusifer* White.

inches thick crowded with casts of '*Platyschisma*' *helicites* (Fig. 40H, I) and *Modiolopsis complanata* (Fig. 40E) as well as ostracods and, towards the top, the small brachiopod *Lingula minima*. Brachiopods characteristic of the underlying Whitcliffe Beds occur in the '*Platyschisma*'-bearing bands in the basal foot or so and fish spines are common in them all. There is much to suggest that the *P. helicites* Beds are the lateral equivalents of the Tilestones of South Central Wales, so well known from the early writings of Murchison.

*Downton Castle Sandstone.*   The basal Downtonian strata are succeeded by 30 to 50 ft (9–15 m) of fine grained, micaceous yellowish-buff sandstone, locally flaggy and current bedded or with shaly layers, but much of it forming a massive building stone which has been extensively worked. West of the Church Stretton Fault the strata correlated with the Downton Castle Sandstone tend to be more silty and contain several greenish shaly layers rich in plant remains, among which are well-preserved early vascular plants (Psilophytales) such as *Cooksonia* and *Zosterophyllum*. Fish fragments are common almost everywhere in the Downton Castle Sandstone and *Lingula minima* is present, particularly in the lower part. At May Hill the Clifford's Mesne Beds, at Usk the Speckled Grit Beds and at Woolhope the lower part of the Rushall Beds are lithologically and faunally comparable with the Downton Castle Sandstone. At Woolhope they contain carbonized remains of *Pachytheca*, a small spherical algal colony, and *Prototaxites*, a probable land plant of unknown affinities.

*Temeside Shales.*   Where separable, mainly in the north, this is a very variable group of greenish rubbly marl-siltstones, greenish shales and micaceous sandstones which pass up into mainly red marls from 50 to 150 ft (15–46 m) above the base. The marly beds locally weather into large spheroidal 'ball-jointed' masses and they give rise to a reddish soil but less conspicuously so than the overlying red marls. Fossils are fairly common and include *Lingula cornea* (Fig. 40G), large ostracods such as *Londinia spp.* and *Hermannina* ("*Leperditia* large species" auctt.), many bivalves and some fish, and eurypterid fragments. At Perton in the Woolhope Inlier, and near Anchor in Clun Forest the curious fossil *Actinophyllum spinosum*, its affinities unknown, has been collected.

*Ledbury Group.*   In terms of thickness by far the greater part of the Downton Series, 1500 ft (457 m) or so, lies above the Temeside Shales but these higher measures are more difficult to subdivide than the lower. They consist of red siltstones (marls) some of them with bands of calcareous nodules (race) or irregular vertical concretions, thin bands of greenish plant and eurypterid bearing shaly mudstone and beds of purplish, highly micaceous sandstone. The best known of the sandstones in the Corve Dale area is the Holdgate Sandstone about 650 ft (198 m) above the Ludlow Bone Bed. Throughout much of the Ledbury Group the sediments are arranged in a roughly cyclical order. A characteristic cycle begins with coarse sandstone with siltstone pebbles and fossil fragments, which is overlain by finer laminated sandstones and siltstones with numerous traces of burrowing animals. These strata pass up into thick unbedded siltstones which locally include bands of calcareous nodules. Such cycles suggest a depositional environment of deltaic mud-flats near the mouth of a large river, which may occasionally have been inundated by the sea but were fairly regularly deluged by river water laden with coarse sediment. In the top 100 ft (30 m) or so of the Ledbury Group in Shropshire, sandstones and conglomeratic cornstones indistinguishable in lithology and fauna from those in the overlying Ditton Series, are common.

Many fish fragments occur in the Ledbury Group, particularly in lenses or pockets in the more pellety sandstones, and two broad faunal zones based on fish remains are recognized; a zone of *Traquairaspis pococki* (Fig. 40B, C) overlain by a zone of *Traquairaspis symondsi*. In addition to *Traquairaspis* the following genera are recorded from this group: *Anglaspis*, *Corvaspis*, *Tesseraspis*, *Ischnacanthus*, *Onchus* and *Kallostrakon*. Among invertebrates, bivalves and *Lingula cornea* are fairly common, and *Hughmilleria* (*Eurypterus*) *pygmaea* (Fig. 40D) and '*Pterygotus problematicus*' are recorded.

## Ditton Series

The middle division of the Lower Old Red Sandstone makes up large areas of the county of Herefordshire, particularly the hilly ground drained by the Monnow in the south-west, and the hills around Bromyard in the north-east. Reference must also be made to the type area around Ditton Priors in south Shropshire just outside the limits of this region. The lowest part of the Ditton Series consists of red siltstones (marls) and sandstones with several thin concretionary limestones (or cornstones), the so-called '*Psammosteus*' Limestones. The individual limestones are laterally impersistent but in most localities the thickest or most conspicuous limestone of the group is called the Main '*Psammosteus*' Limestone and an horizon at or near its base is taken as the base of the Ditton Series. A coarse white sandstone which yields *Traquairaspis symondsi* associated with *Anglaspis sp.* is also a characteristic feature of this part of the succession. At approximately this level the principal faunal change of the Lower Old Red Sandstone takes place, the *Traquairaspis* fauna of the Downtonian being gradually replaced by the *Pteraspis* fauna.

The rest of the Ditton Series (its approximate equivalent in the south being known as the St. Maughan's Group) consists of some 1500 to 2000 ft (457–610 m) of red 'marls' and sandstones with beds of conglomerate, the so-called conglomeratic cornstones, and thin calcareous layers which pass laterally into limestones like the '*Psammosteus*' limestones and are termed 'concretionary cornstones'. The type of rock most characteristic of the series is the conglomeratic cornstone, which is usually a lenticular bed less than 5 ft ($1\frac{1}{2}$ m) thick made up of sub-rounded pebbles of sandstone, calcareous siltstones and limestone in a calcareous sandy matrix. Its base may be erosive on the bed below, but, if no stratum has been removed by erosion the following cyclical arrangement of the sediments is fairly common:

> Concretionary cornstone
> Siltstone with calcareous concretions, particularly near the top, locally fossiliferous
> Siltstone and fine sandstone
> Sandstone, flat-bedded or current-bedded, locally with channel-like structures
> Conglomerate, usually fossiliferous, base commonly erosive.

The environment under which such deposits might have been formed has been likened to that prevailing in the present Colorado delta-floodplain during a single phase of crevassing; the comparison is summarized in Fig. 42.

The cornstone conglomerates yield most of the fish fragments of the Ditton Series, two of the best known forms being *Pteraspis rostrata* (Fig. 40J) and *Cephalaspis whitei*. The following faunal zones are recognized:

> 3.   Zone of *Pteraspis* (*Cymripteraspis*) *leachi*
> 2.   Zone of *Pteraspis* (*Belgicaspis*) *crouchi*
> 1.   Zone of *Pteraspis* (*Simopteraspis*) *leathensis*

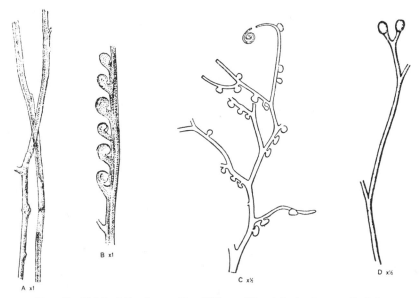

FIG. 41. *Old Red Sandstone Fossil Plants* (*Senni Beds, Brecon Series*)

**A, B.** *Zosterophyllum llanoveranum* Croft and Lang, × 1; **C.** *Gosslingia breconensis* Heard, × ½; **D.** *Cooksonia sp.*, × ½.

### Brecon Series

In this region the highest part of the Lower Old Red Sandstone crops out in the Black Mountains of Brecknockshire, Monmouthshire and Hereford-shire, where it composes all but the summits of fine hills such as Sugar Loaf (Plate XB), Table Mountain, Pen-cerig-calch and all the heights dominating the scenic valleys of the Afon Honddu, Grwyne Fechan and Grwyne Fawr. The same strata also crop out around the Forest of Dean syncline between Monmouth and Mitcheldean. In the Black Mountains the rocks of the Brecon Series are divided into (1) Senni Beds, consisting of a thin basal limestone, the Lower Ffynnon Limestone, followed by some 700 ft (213 m) of pale green to brown micaceous flaggy sandstones with shales and marls with an Upper Ffynnon Limestone, and thin green and red conglomeratic cornstones; and (2) the Brownstones, consisting of 500 to 1000 ft (152–305 m) of reddish brown micaceous sandstones with argillaceous sandstones and brown and green marls. The 'marls' in the Brownstones rarely contain enough calcareous material to form a limestone.

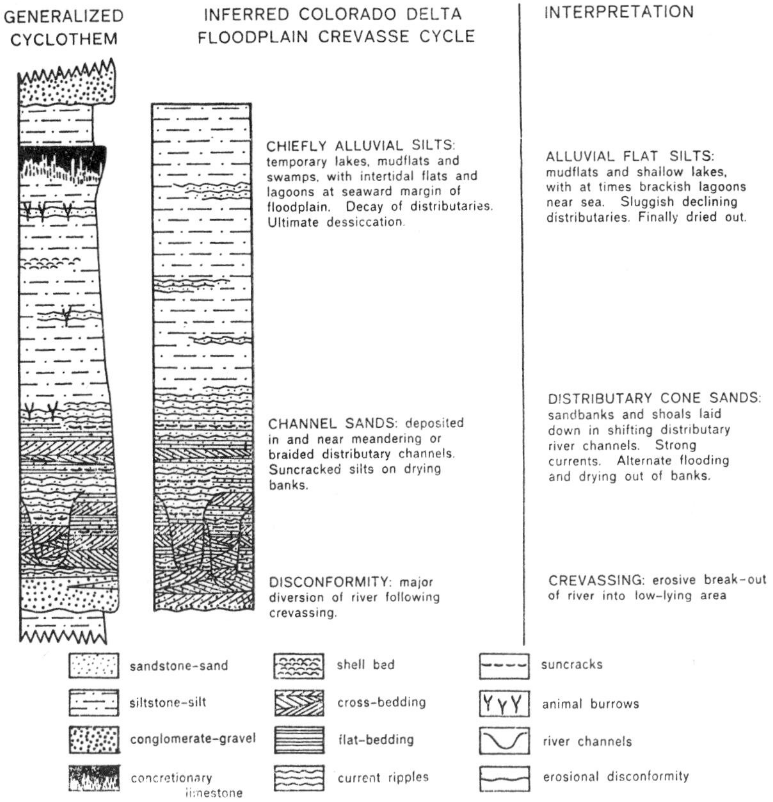

| GENERALIZED CYCLOTHEM | INFERRED COLORADO DELTA FLOODPLAIN CREVASSE CYCLE | INTERPRETATION |
|---|---|---|
| | CHIEFLY ALLUVIAL SILTS: temporary lakes, mudflats and swamps, with intertidal flats and lagoons at seaward margin of floodplain. Decay of distributaries. Ultimate dessiccation. | ALLUVIAL FLAT SILTS: mudflats and shallow lakes, with at times brackish lagoons near sea. Sluggish declining distributaries. Finally dried out. |
| | | DISTRIBUTARY CONE SANDS: sandbanks and shoals laid down in shifting distributary river channels. Strong currents. Alternate flooding and drying out of banks. |
| | CHANNEL SANDS: deposited in and near meandering or braided distributary channels. Suncracked silts on drying banks. | |
| | DISCONFORMITY: major diversion of river following crevassing. | CREVASSING: erosive break-out of river into low-lying area |

| | | | | | |
|---|---|---|---|---|---|
| | sandstone-sand | | shell bed | | suncracks |
| | siltstone-silt | | cross-bedding | | animal burrows |
| | conglomerate-gravel | | flat-bedding | | river channels |
| | concretionary limestone | | current ripples | | erosional disconformity |

FIG. 42. *Interpretation of the cyclic deposits of the Ditton Series* (After Allen and Tarlo.)

East and south of the Black Mountains the Senni Beds are not distinguishable and all the higher part of the Lower Old Red Sandstone is allocated to an undivided Brownstone Group. This Group is probably more than 3000 ft (914 m) thick in the Mitcheldean area. The base of the Brownstone Group south of this region was thought to lie at a somewhat higher horizon than the base of the Senni Beds, but recently a fossil from low in the Group at Mitcheldean, formerly believed to be the Breconian form *Rhinopteraspis dunensis*, has proved to be the Dittonian form *Cymripteraspis leachi*, indicating that here at least part of the Ditton Series has been included in the Brownstone Group.

The Senni Beds at Crickhowell have yielded the true *Rhinopteraspis cornubica* (*dunensis*) as well as numerous plants such as *Drepanophycus*, cf. *Psilophyton princeps*, *Dawsonites arcuatus*, *Zosterophyllum llanoveranum* (Fig. 41A, B) and *Gosslingia breconensis* (Fig. 41c). The Brownstones are almost entirely unfossiliferous.

## Upper Old Red Sandstone

Only three small outliers of Upper Old Red Sandstone occur within this region, all in the extreme south-west, where they form the summits of Sugar Loaf (Plate XB) and Table Mountain (Crucywel) and the adjoining area of high ground around the Carboniferous outlier which caps Pen-cerig-calch. The rocks in these outliers equate with the Quartz Conglomerate formation of the South Wales Coalfield and no break is discernible between them and the Brownstones below.

At Sugar Loaf there is a massive grey grit and at Table Mountain the grit contains many quartz pebbles and closely resembles the Basal Grit of the Millstone Grit. At Pen-cerig-calch the formation is about 200 ft (60 m) thick and it is well exposed in the crag known as the Daren where Murchison recorded the finding of a large scale of *Holoptychius*.

## Igneous Rocks

Small intrusions in the Old Red Sandstone occur at Brockhill on the River Teme north of Shelsley Beauchamp; at Bartestree 3 miles (5 km) east of Hereford; at Great House, 4 miles (6½ km) south-east of Usk, and near Glen Court, Llanllywel, about half-way between Usk and the Great House intrusion.

The Brockhill Dyke is traceable for about ¾ mile (1 km), reaching a maximum thickness of 25 ft (8 m) in Brockhill Quarry, and thinning to about 8 ft (2½ m) on the west bank of the Teme. Where thick the main part is a tesschenite (a doleritic rock with analcite), but an inch or two at the margins have the composition of quartz dolerite. For 30 ft (9 m) or so on either side of the dyke at Brockhill Quarry the Downtonian marls are baked and spotted, and a cornstone band is converted to hornfels.

The Bartestree Dyke is a composite intrusion of a similar nature to the Brockhill Dyke, and the marls and sandstones into which it is intruded are altered in much the same way as at Brockhill.

The intrusion at Great House appears to be a small volcanic vent the known lateral extent of which is more than 150 ft (46 m). The material which occupies the vent is mainly agglomerate consisting of blocks of decomposed monchiquite (a feldspar-free lamprophyre with analcite) with some blocks of limestone and dolomite. Fossils in the limestone blocks indicate that the intrusion is post-basal Carboniferous in age. The north side of the intrusion appears to be cut by a vertical, possibly dyke-like mass of fresh monchiquite which is notable for its very large corroded phenocrysts of pyroxene and other minerals.

The intrusion at Llanllywel is a dyke of monchiquite 15 ft (5 m) wide extending for 20 to 30 yd (18–27 m) in a roughly north-west to south-east direction. It is intruded into strata just below the scarp of the '*Psammosteus*' Limestone.

The age of these intrusions bears comparison with that of contemporaneous igneous rocks in the Lower Carboniferous strata of Weston-super-Mare and the Clee Hills.

# References

ALLEN, J. R. L. 1960. Cornstone. *Geol. Mag.*, **97**, 43–8.

—— 1964. Primary current lineation in the Lower Old Red Sandstone (Devonian), Anglo-Welsh Basin. *Sedimentology*, **3**, 89–108.

—— and TARLO, L. B. 1963. The Downtonian and Dittonian facies of the Welsh Borderland. *Geol. Mag.*, **100**, 129–55.

BALL, H. W., DINELEY, D. L. and WHITE, E. I. 1961. The Old Red Sandstone of Brown Clee Hill and the Adjacent Area. *Bull. Brit. Mus. (Nat. Hist.) Geol.*, **5**, 175–310.

CROFT, W. N. 1953. Breconian: a Stage Name of the Old Red Sandstone. *Geol. Mag.*, **90**, 429–32.

—— and LANG, W. H. 1942. The Lower Devonian Flora of the Senni Beds of Monmouthshire and Breconshire. *Phil. Trans. R. Soc.*, London, (B) **231**, 131–63.

EARP, J. R. 1938. The Higher Silurian rocks of the Kerry District, Montgomery-shire. *Quart. J. Geol. Soc. Lond.*, **94**, 125–60.

ELLES, GERTRUDE L. and SLATER, IDA L. 1906. The Highest Silurian rocks of the Ludlow District. *Quart. J. Geol. Soc. Lond.*, **62**, 195–221.

GREIG, D. C., WRIGHT, J. E., HAINS, B. A. and MITCHELL, G. H. 1968. Geology of the country around Church Stretton, Craven Arms, Wenlock Edge and Brown Clee. *Mem. Geol. Surv.*

HOLLAND, C. H. 1959. The Ludlovian and Downtonian rocks of the Knighton District, Radnorshire. *Quart. J. Geol. Soc. Lond.*, **114**, 449–78.

KING, W. W. 1925. Notes on the "Old Red Sandstone" of Shropshire. *Proc. Geol. Assoc.*, **36**, 383–89.

—— 1934. The Downtonian and Dittonian Strata of Great Britain and North-Western Europe. *Quart. J. Geol. Soc. Lond.*, **90**, 526–70.

LANG, W. H. 1937. On the plant remains from the Downtonian of England and Wales. *Phil. Trans. R. Soc.*, London (B), **227**, 245–91.

MITCHELL, G. H., POCOCK, R. W. and TAYLOR, J. H. 1962. Geology of the country around Droitwich, Abberley and Kidderminster. *Mem. Geol. Surv.*

MURCHISON, R. I. 1839. *The Silurian System.* London.

ROBERTSON, T. 1927. The Geology of the South Wales Coalfield, Pt. 2, Abergavenny, 2nd edit. *Mem. Geol. Surv.*

SQUIRRELL, H. C. and DOWNING, R. A. 1969. Geology of the South Wales Coalfield, Pt. 1. The country around Newport (Mon.), 3rd edit. *Mem. Geol. Surv.*

WELCH, F. B. A. and TROTTER, F. M. 1961. Geology of the country around Monmouth and Chepstow. *Mem. Geol. Surv.*

WHITE, E. I. 1950. The vertebrate faunas of the Lower Old Red Sandstone of the Welsh Borders. *Bull. Brit. Mus. (Nat. Hist.) Geol.*, **1**, 51–67.

# 7. Carboniferous

With the exception of a small outlier of Carboniferous Limestone and Millstone Grit which caps Pen-cerig-calch in the Black Mountains, Carboniferous rocks are represented in this region only by the Upper Coal Measures of the Shrewsbury coalfields and of small areas on and near the Abberley Hills and near Newent in Gloucestershire.

## The Shrewsbury Coalfields

Upper Coal Measures forming what may be termed the Hanwood Coalfield (Fig. 43) crop out in a broad curving tract from the Breidden Hills to Haughmond Hill. They also underlie much of the low ground north-west of the Church Stretton Fault between the slopes of Caer Caradoc and the River Severn, forming what has been called the Leebotwood Coalfield. There is also a small faulted outlier between Dryton and Eaton Constantine. The sequence of the Upper Coal Measures of these coalfields is divisible into three groups which compare with those which seem to persist throughout North Wales and much of the north midlands. These groups are:

> Erbistock Beds (Keele and Enville Beds)
> Coed-yr-Allt Beds (Newcastle Beds)
> Ruabon Marl (Etruria Marl)

They rest with great unconformity on pre-Carboniferous rocks and their general stratigraphical relationship to the Coal Measures of the North Wales (Oswestry) and Coalbrookdale coalfields is shown in Fig. 44.

*Ruabon Marl.* This division is known at outcrop only at the western end of the Hanwood Coalfield, though it may extend, underground, as far east as Great Hanwood. It consists of purple and mottled marls, with some thin bands of impure limestone similar to those that are found in the top of the Ruabon Marl in Denbighshire. This suggests that the attenuation of the marls in the Hanwood Coalfield may be due to overlap of the lower beds.

*Coed-yr-Allt Beds.* These consist of greenish white sandstones alternating with argillaceous beds that include grey and black shales together with red, purple, green and mottled shales and marls. In this group lie the three workable coal-seams of the area. Of these the lowest, or Thin Coal lies about 30 ft (9 m) above the base. The second, or Yard Coal some 65 ft (20 m) higher; and the highest, or Half Yard Coal about 80 ft (24 m) above the Yard. Active working of these seams ceased many years ago (see p. 102).

Between the Yard and Half Yard coals a persistent bed of limestone containing the worm-tube *Spirorbis* is found. This bed is usually in two portions, the lower compact, the upper a breccia of limestone fragments in a calcareous matrix. Near Pontesbury the *Spirorbis* limestone attains a thickness of 7 or 8 ft (2 m), but elsewhere it is 3 ft (1 m) or less in thickness. Near Pitchford the Coed-yr-Allt Beds have a peculiar basement breccia consisting of chips of Longmyndian shale in a sandy matrix, cemented by black 'bitumen'.

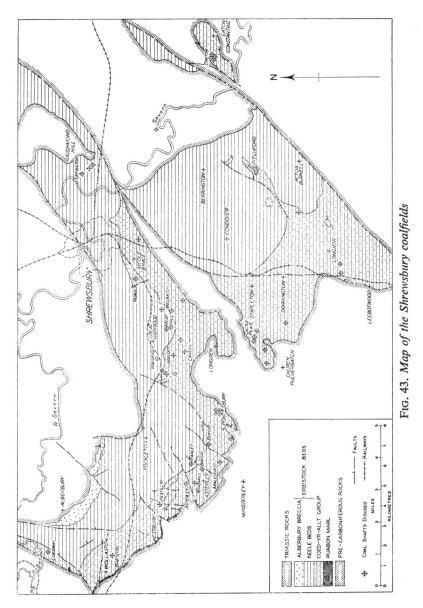

FIG. 43. *Map of the Shrewsbury coalfields*

Natural exposures in the Coed-yr-Allt Beds are rare. Stream sections in Braggington Wood have yielded plant remains and freshwater molluscs; but old mine spoilheaps provide the readiest source of fossils. An interesting fossil that has been found in the Coal Measures of the Dryton Field is *Euproops rotundatus* (Prestwich), an arachnid allied to the kingcrab (*Limulus*).

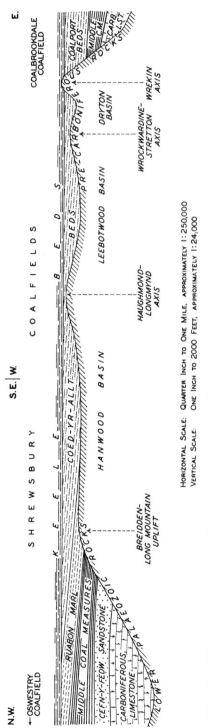

F<small>IG</small>. 44. *Diagrammatic section to illustrate the stratigraphical relation of the Shrewsbury coalfields to the Oswestry and Coalbrookdale coalfields*

*Erbistock Beds.* Most of this highest subdivision of the Upper Coal Measures consists of red marls and sandstones with pellety, calcareous lenses. A limestone containing *Spirorbis* occurs near the base in the Leebotwood Coalfield.

At the top of the group and confined to a narrow outcrop immediately beneath the base of the Triassic Lower Mottled Sandstone, there is a breccia made up of large and small subangular fragments of Carboniferous Limestone, purple sandstone and quartz pebbles in a calcareous matrix. The breccia is interbedded with red marl. This very local formation is called the Alberbury Breccia and is well exposed in Alberbury Park, Rowton Park and in a roadside quarry at Cardeston. It may equate with the Calcareous Conglomerate Group of Central England but the limestone fragments were almost certainly derived from the north-west probably from the neighbourhood of Llanymynech where the Carboniferous Limestone is well developed.

## The Abberley Hills

South-west of Abberley (see map Fig. 46) the southernmost outcrop of the Upper Coal Measures of the Forest of Wyre, folded into a sharp asymmetrical syncline with the eastern limb locally inverted, lies north of Hillside Farm to the west of Woodbury and Walsgrove hills. These strata, which rest unconformably on the Downton Series, are referred to the Highley Beds, a subdivision of the Upper Coal Measures which is lithologically similar to the Halesowen Group of Staffordshire.

A narrow outcrop of Coal Measures extends along the west side of Woodbury Hill, where the measures rest on Upper Ludlow Shales and are overlain, unconformably, by the Clent (Haffield) Breccia. There is a small outcrop of similar beds $\frac{1}{2}$ mile (0·8 km) south-west of Martley, which rest on the Pre-Cambrian; another narrow outcrop extends for $\frac{2}{3}$ mile (1 km) southwards from immediately west of Berrow Farm, and a small outcrop occurs at New Inn, Storridge. The Berrow Coal Measures rest on the Downton Series and are overlain by the Clent Breccia. In the absence of fossils these rocks are all referred to the Highley Beds.

The disposition of the Coal Measures of the Abberley Hills with respect to the older rocks caught up in the main Malvernoid line of disturbance, and its bearing on the orogenic chronology, have been debated at great length. It seems certain that the sharply synclinal Highley Beds north of Hillside Farm are folded to the same degree as the underlying Downtonian, the plane of the unconformity separating the two formations being fully involved. Therefore folding, sufficiently intense to produce local inversion, post-dated the deposition of the Highley Beds.

What is less certain is the precise relationship of the Woodbury Hill Coal Measures to the sharply folded and thrust Silurian rocks beneath them. If the relationship is unconformable and, as the mapping suggests, the plane of the unconformity is relatively undisturbed, then the formation of folds and thrusts in the Silurian pre-dates the deposition of the Highley Beds and belongs to an earlier orogenic phase than the folding north of Hillside Farm. If, on the other hand, the Highley Beds and the Silurian under Woodbury Hill are separated by a thrust or low-angle fault, then there is no need to ascribe the folding of the Silurian to an earlier phase, and all the close fold and thrust structures of the Abberley disturbed belt probably date from a single post-Highley Beds orogenic phase.

## The Newent Coalfield

There are small areas of Upper Coal Measures between the south end of the Malverns and the Silurian inlier of May Hill; the most important out-crop is around Newent. Here the measures consist of grey and reddish mudstones and sandstones and they contain thin coal seams which have been worked at Boulsden south-west of Newent and at Hill House west of Oxen-hall. They have yielded plant remains at Boulsden which indicate an Upper Coal Measures age.

## Clent (Haffield) Breccia

A roughly stratified very coarse breccia which rests unconformably upon Coal Measures and older formations in the Abberley and Malvern Hills is known as the Clent, or Haffield Breccia. Reaching a maximum thickness of 400 to 450 ft (122–137 m) at Osebury Rocks, Knightwick, it consists of angular fragments and blocks of volcanic rocks, usually embedded in a sandy or marly matrix with much hematite staining giving them a dull red colour. Many of the fragments are of volcanic conglomerates, banded rhyolites and porphyrites and some vesicular andesites and melaphyres. Quartzites like those of the Lickey Hills and curious banded shales like those of the Eastern Longmyndian are locally common as well as fragments of fossiliferous Upper Llandovery sandstones, slabs of Wenlock Limestone and, more rarely, limestones of Ludlow age. At Woodbury Hill (Plate IXA) where the Clent Breccia dips gently eastwards, the Lower Keuper Sandstone abuts against it on the north-east side apparently unconformably, while on the west and south sides of the hill the junction with the underlying Highley Beds and Silurian is also unconformable. A road section at Knightwick station is reported to show the breccia passing upwards conformably into normal Bunter Sandstone.

## References

ARBER, E. A. N. 1910. Notes on a collection of fossil plants from the Newent Coalfield (Gloucestershire). *Geol. Mag.*, **7**, 241–4.

LAPWORTH, C. 1898. Sketch of the geology of the Birmingham District. *Proc. Geol. Assoc.*, **15**, 313–389.

MITCHELL, G. H., POCOCK, R. W. and TAYLOR, J. H. 1962. Geology of the country around Droitwich, Abberley and Kidderminster. *Mem. Geol. Surv.*

POCOCK, R. W., WHITEHEAD, T. H., WEDD, C. B. and ROBERTSON, T. 1938. Shrewsbury District including the Hanwood Coalfield. *Mem. Geol. Surv.*

# 8. Triassic

Triassic rocks fringe the region to the north in Shropshire and to the east in Worcestershire. In Shropshire only the Bunter formation is represented within this region and the area which it underlies is largely concealed by superficial deposits. In Worcestershire west of the Severn, large areas are occupied by the Keuper Marl but the underlying Keuper and Bunter sandstones crop out along the Abberley–Malvern–May Hill disturbance.

## Bunter

The Lower Mottled Sandstone of Shropshire is up to 600 ft (183 m) of medium to coarse grained non-pebbly reddish brown sandstone with greenish patches. It is usually highly cross bedded and the steep inclination of the laminae together with the rounded character of the grains suggest that it originated as a wind-borne deposit. Along the flanks of the Uriconian hills there is a basal breccia made up of angular fragments of Uriconian rhyolite and tuff with some Cambrian quartzite.

The Bunter Pebble Beds of Shropshire consist of some 150 ft (46 m) of sandstones of a rather coarser grain than the Lower Mottled Sandstones with pebbly layers and beds of soft conglomerate. The pebbles are mostly of quartz and quartzite but in places they include pebbles of rocks resembling Uriconian rhyolites and tuffs.

The Upper Mottled Sandstone of Shropshire differs little from the Lower Mottled Sandstone. In Worcestershire it forms the base of the Trias alongside the greater part of the Malvern Hills but south of Bromsberrow, to the west of the Malvern Fault, it is overlapped by the Keuper Sandstone.

## Keuper

The Lower Keuper Sandstone of west Worcestershire (Witley, Martley, etc.) consists of several hundreds of feet of coarse, cross-bedded dull red to brown or grey sandstones with scattered pellets of red marl. The basal beds include bands of hard calcareous marl breccia colloquially called 'catbrain', and an exceptionally thick basal breccia at Yarhampton near Abberley Hill is not unlike the Clent Breccia, from which it may be partly derived. Towards the top of the Keuper Sandstone bands of red and green marl are common and there is a transition to the strata which form the Keuper Marl. In an adjacent area of Worcestershire (Rock Hill, Bromsgrove) argillaceous bands in the upper part of the Keuper Sandstone have yielded fossil plants such as *Schizoneura* and *Equisetites* and the early conifer *Voltzia*. Sandstones and marl-conglomerates have also yielded remains of the lung fish *Ceratodus*, present-day relatives of which are able to survive periods of drought. Derived remains of terrestrial animals include the scorpions *Mesophonus* and *Spongiophonus* associated with labyrinthodont amphibians and reptiles. The crustacean *Euestheria* is also common in the beds of mudstone.

The Keuper Marl of west Worcestershire consists of more than 1500 ft
(457 m) of red to reddish brown marls, hard and blocky in character, locally
mottled with green and interbedded with subordinate green marls. Harder
bands are of two types; tough compact red and green marlstones, commonly
dolomitic; and fine grained grey, buff or pale red sandstone bands known as
'skerry' bands. The latter usually consist of quartz with some feldspar, worn
dolomite crystals and other detrital minerals. They are often ripple marked
and show sun-cracks, raindrop pittings and salt pseudomorphs as well as
groove casts, load casts, airheave structures and worm trails. These features
probably indicate the lacustrine origin of the sediment.

In the upper part of the Keuper Marl there is a group of thin grey sandstones and bluish grey shales about 40 ft (12 m) thick and known as the Arden
Sandstone. Elsewhere this group has yielded the remains of plants, vertebrates and *Euestheria*. The well-known evaporite beds of the Keuper Marl
have not been traced within this region though they are present east of the
River Severn and salt is worked at Stoke Prior near Droitwich. The highest
beds of the Keuper Marl consist of a distinctive group, up to 40 ft (12 m)
thick, of grey, green and white marls, which, from their colour, are known as
the Tea Green Marl. They are apparently harder than the red marls which
form the bulk of the Keuper Marl and tend to form slight escarpments.

## Some Post-Keuper Events

Apart from superficial deposits and a small patch of Lower Lias capping
Berrow Hill 3 miles (4·8 km) east of Bromsberrow, no formations later than
the Keuper Marl occur in the Welsh Borderland region and its history in the
remainder of the Secondary and the Tertiary eras can only be inferred from
indirect evidence.

The main outcrops of the Rhaetic and Lias formations approach within
a few miles of the borders of the region in the south, and an outlier of them
lies, at Prees, only some 8 miles (12·8 km) to the north. It is probable that
the seas in which these formations were deposited submerged part of the
region; but they would not appear to have spread much beyond the area
that had been occupied by the lake in which the Keuper Marl was formed.
The sea appears to have extended farther into the region in Middle Jurassic
times, and perhaps submerged the whole of it; but in Upper Jurassic times
it receded, and the whole region may have become dry land again.

In the early part of the Cretaceous period the region was included in an
extensive land area; but, in the later Cretaceous, it became submerged again,
and probably was entirely covered by the sea in which the Chalk was
deposited.

The post-Cretaceous movements elevated the region above sea-level once
more, and as a whole it may never since have been submerged. The present
river system was perhaps initiated upon a surface of slightly flexured Chalk.
The removal of the Cretaceous and Jurassic deposits, however, by laying
bare the uneven surface of the older rocks caused such profound modifica-
tions of the drainage system that only slight suggestions of its original condition remain.

# 9. Structure

The confined outcrops of the pre-Silurian rocks of the Welsh Borderland region and the fact that these older rocks have been subjected to so many periods of major and minor earth movement during the great lapses of time which they span, make it possible to give only a brief outline of their structural features in an essentially local context. In contrast, the Silurian and Upper Palaeozoic rocks crop out extensively over most of the surface of the region and they evidently acquired their main structural features during two fairly well-defined periods of earth movement, those of Middle Devonian (Late Caledonian) and Middle to Late Carboniferous (Armorican) age.

## Structures which pre-date the Late Caledonian Orogeny

The basic structural pattern of south Shropshire appears to have been set during the Pre-Cambrian. It is possible that folds with a north-westerly plunge (Charnoid trend) in the Uriconian of Caer Caradoc and Cardington hills may have resulted from pre-Longmyndian earth movements, but the evidence is not conclusive and they could be volcanic folds of local origin. The thrust-faults on Caer Caradoc are probably also of Pre-Cambrian age (Fig. 9). The major north-north-easterly trending Pontesford–Linley and Church Stretton disturbances appear to have controlled sedimentation in Longmyndian times and have been active at many periods since then. The Church Stretton Fault Complex extends southwards from the Wrekin area to Presteigne, Old Radnor, and beyond, and there is some geophysical evidence to suggest that the Pontesford–Linley disturbance may also have a considerable southward extension under a cover of Silurian rocks. A third sub-parallel fault line appears to lie along the Severn Valley to the west of the Breidden Hills. The Longmyndian rocks between the Pontesford–Linley and Church Stretton disturbances are folded into a major isoclinal syncline with a north-north-easterly axis (Fig. 45). This fold is certainly pre-Caradoc in age, and, as there appear to have been no major earth movements in this area during Cambrian and early Ordovician times, it is inferred that it is of Pre-Cambrian age (Greig and others 1968, p. 277).

The Cambrian rocks of Shropshire show evidence of minor folding and faulting (Fig. 18) but the next earth movements of any magnitude appear to be post-Caradoc and pre-Upper Llandovery in age (early Caledonian or Taconian). At this time the Ordovician rocks of the Breidden Hills were folded into a north-easterly trending anticline with associated minor folds (Fig. 26). The Ordovician rocks of the Shelve area were also folded with the formation of the north-north-easterly trending Shelve Anticline and Ritton Castle Syncline (Fig. 45). Whittard (1952, p. 186) has stated that these rocks were also affected at this time by major northerly trending tear faults and complementary shear faults. The major N.–S. transcurrent faults within the Longmyndian of the Long Mynd (Plate XA) probably also date from this

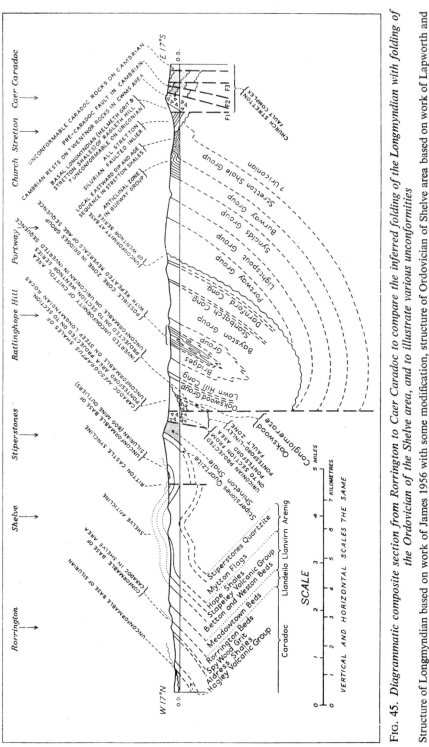

Fig. 45. *Diagrammatic composite section from Rorrington to Caer Caradoc to compare the inferred folding of the Longmyndian with folding of the Ordovician of the Shelve area, and to illustrate various unconformities*

Structure of Longmyndian based on work of James 1956 with some modification, structure of Ordovician of Shelve area based on work of Lapworth and Watts with some modification.

time and there were extensive transverse movements on the north-north-easterly trending faults of the Church Stretton Fault Complex. Immediately to the east, the Cambrian and Ordovician rocks were tilted into their present steeply dipping attitude. All the structural features pre-date the basal Silurian rocks which rest on Ordovician and older strata with strong unconformity.

As well as in Montgomeryshire and south Shropshire, structures of pre-Silurian age occur in the Malvern Hills. The structural history of the Malverns is complex and the pre-Silurian structures have to some extent been modified and obscured by later tectonic events. The Malvernian rocks show evidence of intense tectonism in Pre-Cambrian times (pp. 11, 13), and the overlying Warren House Group appear to have been tilted before the deposition of the basal Cambrian. Structures in the Cambrian have a predominantly north-westerly trend, but, as there is little disconformity between the Cambrian and the overlying Upper Llandovery, it is probable that they are mainly due to post-Silurian movements.

## Late Caledonian and Armorican Structures

The absence of Upper Old Red Sandstone or Lower Carboniferous rocks resting directly on strata older than the Brownstones makes the allocation of structures throughout the Silurian and Old Red Sandstone terrain to either the late Caledonian or Armorican orogeny a matter of conjecture. The trend of the main Silurian outcrop, and of the Church Stretton Fault, as well as the trend of most of the structures in the Silurian and Downtonian rocks of the 'basin' areas, is N.N.E.–S.S.W. (Caledonoid); and, the farther north-west one goes, the more certain it becomes that the structures in these rocks date from the late Caledonian orogeny. South-eastwards there are large areas of near-horizontal or very gently folded Old Red Sandstone and, flanking these areas, are the two complex faulted periclines of Woolhope and Usk; in the extreme south-east there is the great N.–S. disturbance of Abberley–Malvern–May Hill.

If the relationship between the steeply dipping Lower Carboniferous rocks and the unconformably overlying Upper Coal Measures along the east side of the Forest of Dean is taken into consideration, the structures which give rise to these Silurian inliers could be thought to have originated during the same clearly defined period of intra-Carboniferous folding. On the other hand the main movements along the Abberley end of the Malvern 'axis' involve beds as young as the Highley Beds (see p. 85), and therefore occurred during the main (late) Armorican orogeny. In the case of the Usk Inlier, positive movement along a N.–S. axis approximating in position to that of the present Usk Anticline is believed to have been active during mid-Carboniferous times.

Reviewing the Silurian/Old Red Sandstone terrain from north-west to south-east, the Long Mountain Syncline is a broad basin-like structure with its longer axis trending N.E.–S.W. There is minor folding on its north-west limb, but the inclination of its south-east limb is fairly steady, and faulting is small-scale. South-west of the Long Mountain, in the Montgomery–Newtown–Kerry area, there is simple folding along well-defined N.N.E.–S.S.W. axes, and this folded area is adjoined to the south-east by the broad

Felindre Basin. This basin, which occupies almost the whole of the western part of Clun Forest, is a fairly simple and roughly circular structure but it is traversed by N.–S.-trending normal faults with throws of up to about 2000 ft (600 m).

The middle part of Clun Forest and the north-west side of Radnor Forest are traversed by a N.N.E.–S.S.W.-trending belt of fairly sharp folding which locally narrows to a zone of compression. The limbs of the westernmost folds of this disturbed belt are frequently vertical and locally inverted, and, where it transects the Teme and Clun valleys, the western margin of the belt is sharply demarcated by a master fault. Between this major Caledonoid structure and the sub-parallel Church Stretton Fault there is mainly very gentle folding, except for local sharp flexures in the vicinity of the fault zone itself.

The Church Stretton Fault is a complex Caledonoid structure along which the rocks tend to be downthrown thousands of feet to the north-west of a major line of dislocation and sharply anticlinally folded with steep south-eastwardly overthrusting to the south-east. In the Presteigne–Old Radnor area the disturbance consists of two parallel anticlines with normal faulting on the north-west limbs and reverse faulting on the south-east limbs. The association of south-eastwardly overthrust anticlines with normal north-westward downfaulting has been attributed to upthrusting of the pre-Silurian basement along north-east to south-west fractures of pre-Silurian origin.

South-east of the Church Stretton Fault the only important warp in the steadily south-eastwardly dipping Silurian and Downtonian strata of the main outcrop between Buildwas and Kington consists of the Downton (Brown Clee) Syncline and the complementary Ludlow (Ledwyche) Anticline. Apart from these structures the regularity of the outcrop is affected by only a few normal dip faults. The Downton Syncline is an asymmetrical structure which plunges east-north-eastwards; the north limb dips eastwards at about 8° and the south limb dips northwards at about 35°. The Ludlow Anticline trends and plunges in the same direction and the dip on its southern limb is about 10°. Faulting hereabouts is mainly parallel with or perpendicular to the fold axes, but is mainly on a small scale.

From the vicinity of Abberley (Fig. 46) southwards, to near Martley the Malvern 'axis' comprises a N.–S. anticline, the Rodge Hill Anticline, which has been sharply overfolded to the west and north-west. On its western limb a reverse fault, the Cockshot Hill Thrust, has developed, and on its eastern side slices of successively older rocks have been forced westwards along two thrust planes.

South of Martley, though less well exposed, the N.–S. trend of the Malvern 'axis' continues and the Downtonian rocks are anticlinally folded between Ham Farm and Horsham. South of the River Teme, from Ravenshill Wood to Cowleigh Park, an anticlinal structure cored by Llandovery rocks has been called the Storridge or Malvern Anticline. The east limb of this structure is truncated by the Malvern Fault and there is subsidiary folding at Cowleigh Park, but there is little or no surface evidence that in this particular tract the western limb of the anticline is overturned. South of Cowleigh Park the west limb of this anticline continues as the west limb of the main Malvern Anticline at least as far as Herefordshire Beacon. Of the strata caught up in this

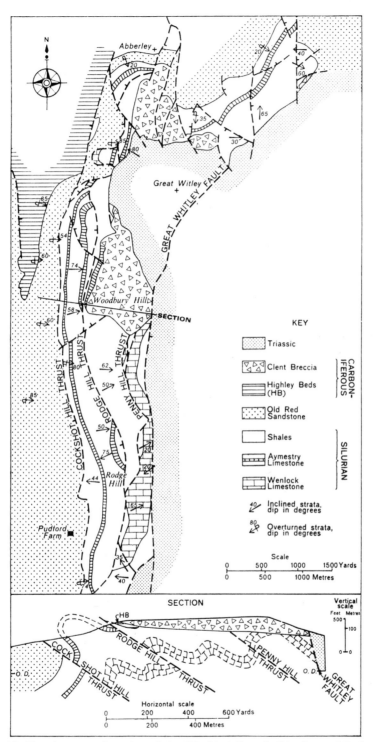

FIG. 46. *Sketch-map and section of the Silurian inlier of Abberley and Rodge Hill*

structure those closest to the Pre-Cambrian core of the fold are almost every-where inverted. In the north only the lowest Llandovery beds are inverted but southwards the inverted part of the limb is wider and, near Herefordshire Beacon, all the strata are caught up in a structure overturned towards the west probably with thrusting. Near Colwall station there are N.W.–S.E. faults the most conspicuous of which throws the Wenlock Limestone against Downtonian near the railway tunnel mouth at Colwall.

West of the main Malvern 'axis' there is parallel subsidiary folding between Worcestershire Beacon and Mathon, west of Herefordshire Beacon around Chances Pitch, and west of Swinyard Hill. The Malvern folds are separated from those of Ledbury by the Eastnor Syncline, a N.–S.-trending structure with varying plunge. The Ledbury folds are a series of mainly N.N.E.–S.S.W.-trending periclines and associated synclines, and the Wenlock rocks in the core of the inlier around Coneygree and May Hill woods display a dome-like structure with subsidiary folds along north-easterly axes. To the west all these structures run against the N.–S. Ledbury Fault, a major down-throw west, but it is uncertain to what extent they continue into the Old Red Sandstone west of the fault.

The structure of the Woolhope Inlier is an asymmetrical anticline trending from N.W.–S.E., with a steeper western limb, complicated by subsidiary dome structures at Haugh Wood and Broadmoor Common separated by the Rudge End Fault. Faults are mainly N.E.–S.W. or N.–S. wrench faults and there is a western boundary fault, a reverse fracture, the plane of which has been dislocated by the wrench faults. The small Shucknall Inlier to the north is a N.W.–S.E. asymmetrical anticline with a steeper eastern limb. It has been postulated that the Bartestree Gap which separates the Shucknall and Woolhope structures is occupied by a disturbed belt which may extend south-westwards, via Pontrilas and Pandy, and between Sugar Loaf and Table mountains to join the Vale of Neath Disturbance.

Situated at the convergence of the Woolhope and Malvern axes of folding the May Hill Inlier is a complex structure in which the main north-north-westwardly trending 'May Hill' Pericline gives place southwards to the N.–S.-trending 'Huntley Hill' Pericline. The north part of the main pericline is closed by a northwardly pitching anticline flanked to the east by a very shallow syncline. In the extreme south the beds on the extension of the west limb of the Huntley Hill Pericline are inverted. A major fault downthrowing to the east defines the south-eastern limit of the inlier and another, down-throwing to the west defines much of the north-western boundary of the inlier. The northern half of the inlier is traversed by a N.W.–S.E. master fault on either side of which the fold axes do not match.

The structure of the Usk Inlier is a pericline with its major axis trending slightly east of north. Dips on the two major limbs average about 10°; the southward pitch is at about 7° and, to the north, the pitch appears to be similar but with rather steeper dips on the western flank. Between Graigwith House and Usk the eastern limb of the main pericline is complicated by a subsidiary syncline and an anticline, both asymmetrical with steeper eastern limbs, the axes of which trend from N.N.E.–S.S.W. Both axes are displaced by a series of faults trending from N.W.–S.E. These faults are probably normal faults and there is a fairly complex pattern of faults on this side of the inlier. Faults on the west side of the inlier are fewer and are mostly normal faults which trend roughly parallel with the main fold axis.

## Post-Triassic Structures

Evidence of post-Triassic earth movements in the Welsh Borderland region is provided by the gently tilted lie of the Bunter Sandstone in the Shrewsbury area and the transection of the Bunter outcrop by a number of mainly N.E.–S.W. faults. The Church Stretton Fault throws Bunter Sandstone against Uriconian rocks at Eyton-on-Severn and Wrockwardine and the Ercall Mill Fault throws Bunter Sandstone against Erbistock Beds. The Triassic rocks of west Worcestershire are also gently folded and broken by a few faults, and the eastern boundary fault of the Ledbury–Malvern and May Hill inliers may be partly post-Triassic in age.

# 10. Pleistocene and Recent Deposits

Little is known of the history of the Welsh Borderland region during the greater part of the Pleistocene Period when southern and eastern England were subjected to complex erosional and depositional cycles dominated by the earlier Pleistocene glaciations. There is, however, abundant evidence that during the last 100 000 years or so the region was subjected to very severe glacial conditions. It is possible that a main and extremely rigorous glaciation prior to about 60 000 years b.p. was followed by a prolonged period of deglaciation which cleared most of the ice from the region, and that this was followed, possibly between about 30 000 and 20 000 years b.p., by a considerable readvance of the ice. However, only in central Shropshire to the north and in those parts of the Severn Valley which fall within the eastern limits of the region is there evidence for other than a monoglacial interpretation of the Pleistocene deposits of this region.

## The Advance of the Ice

During the main glaciation most of the region was invaded by streams of ice fed from the mountains of Wales to the west and north, while the Shropshire area was overwhelmed by the south-eastern lobe of a great sheet of ice fed from west Scotland and the Lake District which filled the Irish Sea basin. With the development of glacial conditions the Irish Sea ice extended from the Shropshire lowlands southwards into Ape Dale and up the Church Stretton Valley, while ice from the north-west moved into the head of the East Onny Valley. Possibly at the same time Welsh ice filled the Camlad Valley and advanced around the southern end of the Long Mynd, across the low ground of the Wistanstow area and invaded the southern part of Ape Dale. This ice was probably confluent with a glacier issuing eastwards from the Clun Valley, and the combined stream apparently advanced eastwards through the valleys near Norton Camp into the lower part of Corve Dale. It is uncertain whether the maxima of the Welsh and Irish Sea ice occurred at the same time.

The movement of the ice in the southern part of the district was, in detail, somewhat complex. The high ground in the west of the district tended to deflect the ice coming from Central Wales. The upper parts of the Clun and Teme valleys contain drift of local origin only; but the Welsh ice escaped north-eastward, over the col on which the sources of the rivers Clun and Mule now lie, into the valleys of the Mule and Caebitra. It reached a height of at least 1750 ft (533 m) above Ordnance Datum on Radnor Forest, and left boulders of dolerite like that of Baxter's Bank, 9 miles (14·4 km) to the north-west, at a height of 1400 ft (427 m) on the slopes of Bryn-y-Main. In places it escaped eastward, through the valley of the River Lugg, to the north of Radnor Forest, and over the cols in the high ground farther south. In general, however, the ice-stream was deflected southward and south-eastward to reinforce the great glaciers of the Wye and Usk valleys.

The Wye glacier, emerging from the narrow valley to the north-west of the Black Mountains, and augmented by ice passing over cols in the high ground to the west, spread out in a great lobe over the Herefordshire lowland. This lobe, with the masses of morainic and outwash drift that it left banked against the hills from Kington to Orleton, exercised a great influence upon the drainage of the area. Boulders recognizable as from Hanter Hill and Stanner Rocks were carried as far east as Crick's Green, 3 miles (5 km) south-south-west of Bromyard. During its retreat the ice of the Wye glacier left terminal moraines across the valley in places above Hereford. An interesting deposit in the valley of the Wye at Bredwardine and Breinton is a grey clay containing foraminifera. It appears to be older than at least part of the Glacial deposits, and presumably indicates an inlet of the sea at some time before the maximum extension of the ice.

The Wye glacier did not override the Black Mountains, the valleys of which contain only boulders of local origin.

The Usk glacier has left morainic deposits and outwash gravels over a considerable area near Abergavenny.

The boulder-clay deposited by the Irish Sea ice is commonly reddish in colour and, in many places, sandy. That from the Welsh ice is generally brown; but where the ice passed over red rocks, such as the Keele Beds, it is red. Boulders of granite and other rocks from Scotland and the Lake District have been left at a height of 800 ft (244 m) above sea-level on Wenlock Edge, and at over 1000 ft (305 m) near Woolstaston and Pulverbatch; whilst boulders of Stiperstones Quartzite lie at over 1300 ft (396 m) on Wilderley Hill near Smethcott. Glacial striae have been noted on Charlton Hill and on Sharpstone Hill, indicating ice-movements from north-north-west to south-south-east and from north to south respectively.

## Glacial Retreat Phenomena

The melting and retreat of the ice of the main glaciation from the Welsh Borderland region released enormous quantities of water into an area many of the former drainage routes of which were blocked by stagnant ice or choked with glacial debris. There were many areas where the drainage was so effectively dammed that great volumes of impounded static water, much of it sub-glacial, accumulated, while escaping surface and sub-glacial waters cut channels into everything which impeded their flow, and many of the channels transected former watersheds. Erosive and depositional activity reached a crescendo of vigour as channels were deepened and vast masses of till were swept away by the torrents; glacial sand and gravel were deposited within and around the melting ice and fluvio-glacial terraces of sand and gravel accumulated rapidly downstream in valleys emerging from a rigorous peri-glacial regime. Clay was mostly carried away by the swollen waters but some of it accumulated as laminated clays in temporary lacustrine areas.

An area where many abandoned melt-water channels may be seen is on the eastern slopes of the Long Mynd to the north and west of All Stretton (Plate XIA). They occur at various levels between about 1050 ft (320 m) O.D. and 600 ft (183 m) O.D. and in some instances long spurs are incised at

several levels in an eastward-descending sequence. At an early stage in the
retreat melt water from an elevated ice front across the north-east shoulder
of the Long Mynd would have flowed southward towards Church Stretton
across the ice-filled Long Mynd batches cutting valleys in the intervening
spurs. With the gradual northward withdrawal of the Church Stretton glacier,
the ice in the tributary valleys would sink to progressively lower levels
allowing water from some of the channels to follow the courses of these
partly vacated valleys down to the main valley and on southwards to a
final escape via the Worldsend channel at Church Stretton.

The present course of the River Onny, via the Plowden channel and round
the south end of the Long Mynd, was probably established when water from
melting ice occupying the broad valley to the north-west grew in volume. Its
access to the Kemp Valley, a more natural outlet to the south, must have
been blocked. After the Plowden channel was cut back beyond and below
the lip of the broad valley west of the Long Mynd, drainage from the ground
north and west of Plowden would continue to excavate the new course. At
a later stage part of the Camlad Valley farther west may have harboured a
lake, held back by ice in the Severn Valley to the north. The escaping water
excavated the gorge known as Marrington Dingle, a channel which falls
northwards towards the broad Marton Valley between the Long Mountain
and Shelve hills, suggesting that the Marton Valley and its north-eastern
exit were then clear of ice. Farther west again the River Mule, which formerly
flowed in the broad valley now drained by the Caebitra, was probably
diverted by great masses of ice and glacial debris at Sarn, and it cut a remarkable gorge northwards through the hills separating its valley from that
of the Severn.

The most spectacular new water course to evolve from the chaos of deglaciation was that now known as the Ironbridge Gorge. In pre-glacial times
a continuous watershed probably extended across the site of the present
Severn Valley near Ironbridge. The drainage of the area to the north-west
of this watershed flowed northward into the Irish Sea. This drainage pattern
was disrupted when glacially impounded water occupying what is now Coalbrookdale overflowed southwards across the watershed into the head of a
tributary of the River Stour. With a retreat of the ice-front and a lowering
of the level of overflow across the newly breached watershed the Coalbrookdale 'lake' became merged with a larger one covering the site of Buildwas. The further retreat of the ice sheet enlarged the area of water impounded
between the ice front and the former watershed until all the low ground lay
beneath water the level of which appears to have stood for a prolonged
period at about 300 ft (91 m) O.D. Into water at about this level very large
areas of the so-called Middle Sands of Shropshire were spread, and it is to
this long-lived expanse of ice-dammed water that the name 'Glacier Lake
Lapworth' is most aptly applied. The name was given in honour of the
geologist who first suggested the now generally accepted explanation of the
origin of the Ironbridge Gorge.

It has long been suspected that the great terrace which flanks the River
Severn south of the Ironbridge Gorge, named by Wills the Main Terrace,
was formed contemporaneously with the 'Lake Lapworth' retreat phase north
of the gorge, and recent radiocarbon dates suggest that this supposition is

**Plate X**

A. Aerial view along the western scarp slope of the Long Mynd (*Cambridge University Aerial Photograph No. EN58*)

(*For full explanation see p. xi*)

B. The Sugar Loaf near Abergavenny (A 3282)

A. Glacial drainage channel, Cwmdale, All Stretton                    (A 9431)

*(For full explanation see p. xi)*                                    **Plate XI**

B. Barite vein in Western Longmyndian, Wrentnall Mine                 (A 4828)

true if the name 'Lake Lapworth' is applied to the retreat phase which preceded the formation of the so-called Upper Boulder Clay.

Almost as spectacular, but on a smaller scale than those of the Severn, were the changes brought about in the course of the River Teme below where it emerges from its mountain confines at Brampton Bryan. From the low ground around Leintwardine it probably flowed northwards via Clungunford and Sibdon Carwood to pass through what was probably the only pre-glacial breach in the long barrier escarpment of Ludlow rocks, the valley from Craven Arms to Onibury. With access to this outlet blocked it appears to have flooded whatever part of the Vale of Wigmore was free of stagnant ice until it spilled eastwards to carve itself a new channel through the escarpment of Ludlow rocks, the present Downton Gorge. South of Ludlow it probably followed the main valley southwards to Leominster, the present Tenbury Valley being that of a westwardly flowing tributary. Ice and morainic material at Orleton effectively severed its route from Ludlow to Leominster and the resultant flood conditions appear to have breached the former col at Knightsford Bridge. Downcutting of the col established the Teme firmly in the Tenbury Valley as an eastwardly flowing stream. In this part of its course from Woofferton to Knightsford Bridge the profile and tributaries of the Teme bear clear evidence of its remarkable late Pleistocene history.

The present course of the Wye, in particular its deeply incised meandering course south of Hereford is undoubtedly different in some respects from its pre-glacial course, but its history has been too complex (Miller 1935) for glacial effects to be easily distinguished from earlier influences. Its tributary the River Troddi which formerly probably flowed to join the Usk via the valley of Olway Brook was apparently diverted at Dingestow by ice of the Usk glacier and it now flows eastwards through the gap at Mitchell Troy to join the Wye at Monmouth.

By about 35 000 to 30 000 years b.p., although ice fields probably persisted in northern Britain and in North Wales, the whole of the Welsh Borderland region was probably deglaciated. Lake Lapworth was drained, and vast terraces of glacial sand and gravel covered the Shropshire plain. There followed what appears to have been a cold phase, which may have lasted for 10 000 to 15 000 years, when ice from both North Britain and North Wales was reactivated and moved again into the northern margins of this region. It deposited thin but extensive sheets of till (the so-called Upper Boulder Clay) in central Shropshire above the thick sands and gravels of the earlier glaciation, but throughout most of the region there is little evidence of these renewed glacial conditions.

The melting and retreat of the ice of the Upper Boulder Clay glaciation in the Shropshire/Cheshire plain has been the subject of much recent controversy. It is likely that the area between the Ironbridge Gorge and the retreating ice front suffered inundation by melt water to a limited extent but, if the earlier retreat had cut down the gorge to below 300 ft (91 m) O.D. it is difficult to envisage a repeat of the Lake Lapworth story. The existence, in parts of Shropshire of lake shoreline features cut into Upper Boulder Clay at levels of 250 ft (76 m) O.D. and below has been postulated. These features may well provide a useful record of the retreat of the Upper Boulder Clay ice. Downstream in the Severn Valley, if, as now seems probable, the Main Terrace was deposited at the time of the 'Middle Sands' retreat, the melting of the ice of the Upper Boulder Clay may have been contemporaneous with the formation of the Worcester Terrace.

## Post-Glacial

Evidence of the geological history following the final deglaciation of the Welsh Borderland region is furnished by deposits of peat in waterlogged hollows and on the more elevated tracts of moorland, and by floodplain flats and low-lying terraces of silt, sand and gravel along the floors of the river valleys. The higher terraces in the river valleys mostly date from some period during the last glaciation or earlier.

The late-Glacial and post-Glacial peat sequence in Britain is now well known and both periods have been divided into a number of zones based on the pollen content of the peat. Approximate absolute dates of the zones have been deduced by radiocarbon analyses of carbonaceous material.

Zone I represents the earliest organic material to have been deposited on the Drift of the so-called Scottish Readvance; Zone II is considered to represent a brief warm phase and Zone III suggests a phase of colder climate which is believed to have been contemporaneous with the so-called Highland Readvance in Scotland. Zones I to III are grouped as late-Glacial and the division between the late-Glacial and the post-Glacial is taken at the change from Zone III (Upper Dryas) peat to Zone IV (pre-Boreal) peat. This change occurred approximately 10 000 years ago. Following the pre-Boreal (Zones IV and V) phase the later peats show progression from 'Boreal', through 'Atlantic' to 'sub-Boreal' climatic phases, the last mentioned extending into the Bronze Age of the archaeological chronology.

As neither the ice of the Highland Readvance nor that of the Scottish Readvance reached the Welsh Borderland region, one might expect that peat much older than Zone I would occur in self-enclosed hollows in this region. However, none has yet been recorded, and, at the nearest closely studied site, near Penkridge in Staffordshire, the peat sequence begins with organic material no older than about 10 500 years b.p. At Church Stretton plant material from Late Glacial clay at a depth of 40 ft (12 m) has been dated at about 11 000 years b.p. The moorland peat has not been much studied but, by analogy with other moorland areas, it probably accumulated mainly within the last 4000 years or so.

A superficial deposit that deserves special mention is the great mass of calcareous tufa, or travertine, one of the largest in the country, that forms the Southstone Rock on the south-west side of the Teme Valley, north-west of Shelsley Walsh. This is a cliff over 50 ft (15 m) high and the tufa of which it is formed has been quarried for building stone.

# References

DWERRYHOUSE, A. R. and MILLER, A. A. 1930. The Glaciation of Clun Forest, Radnor Forest and some adjoining districts. *Quart. J. Geol. Soc. Lond.*, **86**, 96–129.

GREIG, D. C., WRIGHT, J. E., HAINS, B. A. and MITCHELL, G. H. 1968. Geology of the country around Church Stretton, Craven Arms, Wenlock Edge and Brown Clee. *Mem. Geol. Surv.*

MILLER, A. A. 1935. Entrenched meanders of the Herefordshire Wye. *Geogr. Journ.*, **85**, 160–78.

MITCHELL, G. H., POCOCK, R. W. and TAYLOR, J. H. 1962. Geology of the country around Droitwich, Abberley and Kidderminster. *Mem. Geol. Surv.*

POCOCK, R. W., WHITEHEAD, T. H., WEDD, C. B. and ROBERTSON, T. 1938. Shrewsbury District including the Hanwood Coalfield. *Mem. Geol. Surv.*

POOLE, E. G. and WHITEMAN, A. J. 1961. The glacial drifts of the southern part of the Shropshire-Cheshire basin. *Quart. J. Geol. Soc. Lond.*, **117**, 91–130.

WILLS, L. J. 1924. The development of the Severn valley in the neighbourhood of Ironbridge and Bridgnorth. *Quart. J. Geol. Soc. Lond.*, **80**, 274–314.

—— 1938. The Pleistocene development of the Severn from Bridgnorth to the sea. *Quart. J. Geol. Soc. Lond.*, **94**, 161–242.

# 11.  Economic Geology

The economy of the Welsh Borderland region is primarily based on agriculture and stock rearing and the principal towns retain their ancient role as agricultural market towns for the surrounding rural areas. Only Shrewsbury and Hereford have, in the last few decades, developed manufacturing industries on a moderate scale.

Exploitation of minerals is far less extensive today than it was in the past when coal was mined in the Shrewsbury coalfields, lead and other vein minerals exploited in the Shelve country, limestone quarried and burned in numerous places, local clays dug for brick-making, and stone, in great variety, hewn for freestone, ashlar and tiles for building or crushed for the surfacing of roads. At present, mineral-based industries are limited to the quarrying of igneous and other hard rocks for road metal and crushed aggregate, mainly in Shropshire and the Malvern Hills; the quarrying of the thicker Salopian limestones for road metal and lime mainly near Much Wenlock, Old Radnor and Woodbury Hill; and the winning of sand and gravel in places such as the Shrewsbury–Buildwas area of the Severn Valley and the Drift-covered low ground west and north of Hereford. There are also small brick and tile works in a few places.

## Coal

In the Shrewsbury coalfields the workable coals lie within the Upper Coal Measures Coed-yr-Allt Beds; only three seams were exploited. The lowest seam, called the Thin or Deep Coal is 1¼ to 1¾ ft (0·38–0·53 m) thick and, being of better quality than the higher seams, was the one most extensively worked. The Yard Coal, lying some 50 to 80 ft (15–24 m) above the Thin Coal, varies from 2½ to 3 ft (0·76–0·91 m) in thickness, but was locally in two leaves. The Half Yard Coal, about 70 to 100 ft (21–30 m) above the Yard Coal reaches a thickness of 1 ft 3 in (0·38 m). The maximum depth of the workings was approximately 500 ft (150 m) and the last important pits, operated by the Hanwood and Moat Hall Colliery Company, closed in 1941. A small colliery, Castle Place Colliery near Pontesbury, was worked until 1947.

## Vein Minerals

Faults and fractures which cut anticlinally folded Mytton Flags around the village of Shelve and similar structures which cut the main outcrop of the Mytton Flags between Minsterley and Ritton Castle carry mineral veins in which the sulphides are mainly galena and sphalerite and the gangue mainly barite and calcite. Some of these veins were rich in galena and this ore was worked from Roman times until the early part of the present century. Towards the end of this period a little sphalerite and barite were also mined. The best known mines were Roman Gravels at Shelve and Snailbeach south of Minsterley.

West, south-west and south of Shelve, faults mainly cutting Ordovician volcanic rocks overlying the Hope Shales are mineralized with small quantities of sulphide in a mainly barite gangue. Barite was mined from these veins mainly between 1850 and 1914, the most important mine being Wotherton, south-west of Meadowtown. East of Shelve a series of mainly E.–W. faults cutting the Longmyndian grits and sandstones south of Habberley are mineralized by barite, and some contain small quantities of the secondary copper minerals, chalcocite and malachite. Barite and very small quantities of copper ore have been worked from these veins. Of the two most important mines, Wrentnall (Plate XIB) was in production from 1890 to 1925 and Huglith, which produced nearly 300 000 tons of barite, was worked from 1910 to 1945. Very small amounts of barite have also been worked along faults cutting Ordovician and Silurian rocks south-east of Breidden Hill.

## Quarries

It is probably true to state that in the Welsh Borderland region almost every rock type reasonably resistant to weathering or abrasion has been quarried locally for building, walling or road surfacing while the purer limestones have, in addition, been quarried for lime. In the case of most of the innumerable old quarries, unless the rock is clearly of a 'tilestone' quality, or the remains of a limekiln are present, it is not now possible to determine for what use a particular quarry was excavated.

Probably the two stratified formations most extensively exploited wherever they crop out were (1) the Wenlock Limestone which could provide roadstone, building stone and the material for lime, cement and smelting flux, and (2) the Tilestones–Downton Castle Sandstone horizon which provided some of the best roofing tiles as well as ashlar and massive freestone.

At the present time quarrying is mostly for limestone products, road metal and concrete aggregate. There are active or only recently suspended quarries for limestone products such as agricultural lime and flux, and a certain amount of roadstone, in the Woolhope Limestone near Old Radnor, in the Wenlock Limestone near Much Wenlock, Wigmore, Presteigne, Mathon (Malvern) and Rodge Hill (south of Woodbury Hill), and in the Aymestry Limestone at Woodbury Hill and Perton near Stoke Edith. Crushed aggregate and most of the roadstone produced in the region come from igneous intrusions and the harder Ordovician, Cambrian and Pre-Cambrian rocks. Ordovician siltstones are quarried for this purpose at Callow Hill, Minsterley, the Ordovician Stiperstones Quartzite at Nills Hill, Pontesbury and the similar Cambrian Wrekin Quartzite at the Ercall Quarry, Wellington. Pre-Cambrian sandstones are quarried near Old Radnor and various igneous and metamorphic rocks in the Malvern Hills. Igneous rocks of various ages are, or have recently been, exploited at Haughmond Hill, Leaton, Bayston Hill, Maddocks Hill, Breidden Hill and Squilver in Shropshire and at Buttington in Montgomeryshire.

The purple shales of the Llandovery Series are excavated for the manufacture of bricks and tiles at Buttington and shales in the Ditton Series are wrought near Bromyard.

**Water Supply**

The demand for water, though steadily increasing, is much less than in the industrial Midlands to the east. Despite the moderate relief and rainfall, few towns draw their supplies from impounding reservoirs. An exception is Church Stretton, supplied from a small reservoir on the Long Mynd. The larger towns draw most of their water from the rivers. Thus Shrewsbury and Worcester obtain their supplies from the Severn and Hereford from the Wye. The potential abstraction from the River Severn has recently been increased by the building of a regulating reservoir on one of its tributaries, the Afon Clywedog near Llanidloes in western Montgomeryshire.

Local supplies are obtained from springs and underground sources throughout the region. The springs located along the boundary faults on either side of the Malvern Hills have long been famed for their purity. Water percolating through the broken and fissured Pre-Cambrian rocks is thrown out where the Pre-Cambrian is faulted against impermeable Palaeozoic and Triassic strata. The purest, softest water, however, comes from springs within the Pre-Cambrian outcrop such as St. Ann's Well, above Great Malvern.

Small amounts of water, usually rather hard, are obtained from the Silurian limestones, and Ledbury gets some of its supplies from this source. In the Old Red Sandstone the Downton Castle Sandstone yields generally small amounts, as at Kington, but the main aquifer is the group of sandstones above the '*Psammosteus*' Limestone. Supplies are obtained from boreholes in these Dittonian sandstones, as at Bromyard and Ross-on-Wye, and also from the spring line where they rest on the impermeable Downtonian marls. Some places on the margin of the region obtain their supplies from the Coal Measures or Triassic rocks.

Drifts and river gravels yield moderate amounts of water in many areas. Leominster is supplied from the gravels of the River Lugg, Craven Arms from the River Onny, and supplies are also obtained from the gravels of the River Clun at Clungunford.

# References

ANON. 1960. River Severn Basin Hydrological Survey. Hydrometric Area No. 54. *Min. Hsing and Local Govt.* H.M.S.O. London.

ANON. 1965. River Wye Basin Hydrological Survey. Hydrometric Area No. 55. *Min. Hsing and Local Govt.* H.M.S.O. London. 57 pp.

DINES, H. G. 1958. The West Shropshire Mining Region. *Bull. Geol. Surv. Gt Brit.* No. 14, 1–43.

POCOCK, R. W., WHITEHEAD, T. H., WEDD, C. B. and ROBERTSON, T. 1938. Shrewsbury District including the Hanwood Coalfield. *Mem. Geol. Surv.*

RICHARDSON, L. 1935. Wells and Springs of Herefordshire. *Mem. Geol. Surv.*

SMITH, B. and DEWEY, H. 1922. Lead and Zinc Ores in the Pre-Carboniferous Rocks of West Shropshire and North Wales. *Mem. Geol. Surv. Min. Resources*, **23**.

# 12. Geological Survey Maps and Memoirs dealing with the Welsh Borderland Region

## Maps

(a) **On the scale of 4 miles to 1 inch (1/253,440)**

Colour-printed; Solid Edition; Out of Print

Sheet 9. (with 10). North Wales and Borders.
Sheet 11. North Midlands.
Sheet 14. South-Central Wales and Borders.
Sheet 16. South Midlands.
Sheet 18. Areas adjoining Bristol Channel.

(b) **On the scale of 1 mile to 1 inch (1/63,360)**

(i) The New Series Sheets are colour printed; some are printed in separate 'solid' and 'drift' editions, others in one 'solid and drift' edition.

152 (Shrewsbury); 166 (Church Stretton); 182 (Droitwich);
232 (Abergavenny); 233 (Monmouth); 249 (Newport);
250 (Chepstow).

(ii) The Old Series Sheets are hand coloured and are not now in print. The following should be consulted for areas where no New Series Sheets have been published:

42 N.E. (North Black Mountains); 42 S.E. (South Black Mountains); 43 N.W. (Hereford); 43 N.E. (South Malverns); 43 S.W. (Ross-on-Wye); 43 S.E. (May Hill); 55 N.W. (Ludlow); 55 N.E. (Cleobury Mortimer); 55 S.W. (Leominster); 55 S.E. (North Malverns); 56 N.E. (Upper Teme, Radnor Forest); 56 S.E. (Old Radnor); 60 N.E. (Long Mountain, Breidden Hills); 60 S.E. (Clun Forest, Shelve).

(c) **On the scale of about $2\frac{1}{2}$ inches to 1 mile (1/25,000)**

SO 48 (Craven Arms); SO 49 (Church Stretton);
SO 59 (Wenlock Edge).

(d) **On the scale of 6 inches to 1 mile (1/10,560)**

The area covered by New Series one-inch scale maps is also covered by maps on the six-inch scale. Where these maps include Coal Measures they are mostly published; for other areas they are deposited for reference in the Library of the Institute of Geological Sciences, Exhibition

Road, London S.W.7. Uncoloured photo-copies may be supplied on special order.

(e) **Geophysical Maps**

    (i)   **On the scale of 4 miles to 1 inch (1/253,440)**

        Gravity Survey Overlay
        Sheet 15. Includes Worcestershire and parts of Shropshire and Herefordshire.

    (ii)   **On the scale of about 4 miles to 1 inch (1/250,000)**

        Aeromagnetic
        Sheet 5. English Midlands and Welsh Borders.

## Memoirs and Sheet Explanations

GREIG, D. C., WRIGHT, J. E., HAINS, B. A. and MITCHELL, G. H. 1968. Geology of the Country around Church Stretton, Craven Arms, Wenlock Edge and Brown Clee (Sheet 166).

HAINS, B. A. 1969. The Geology of the Craven Arms area (Explanation of 1:25 000 Geological Sheet SO 48).

— 1970. The Geology of the Wenlock Edge area (Explanation of 1:25 000 Geological Sheet SO 59).

MITCHELL, G. H., POCOCK, R. W. and TAYLOR, J. H. 1962. Geology of the Country around Droitwich, Abberley and Kidderminster (Sheet 182).

POCOCK, R. W., WHITEHEAD, T. H., WEDD, C. B. and ROBERTSON, T. 1938. Shrewsbury District including the Hanwood Coalfield (Sheet 152).

ROBERTSON, T. 1927. The Geology of the South Wales Coalfield, Part II, Abergavenny, Second Edition (Sheet 232).

SQUIRRELL, H. C. and DOWNING, R. A. 1969. Geology of the South Wales Coalfield, Part I, The Country around Newport (Mon.), Third Edition (Sheet 249).

WELCH, F. B. A. and TROTTER, F. M. 1961. Geology of the Country around Monmouth and Chepstow (Sheets 233 and 250).

WRIGHT, J. E. 1968. The Geology of the Church Stretton area (Explanation of 1:25,000 Geological Sheet SO 49).

# Index

107

Printed in England for Her Majesty's Stationery Office by Penshurst Press, Longfield Road, Tunbridge Wells, Kent.
Dd 717063 C150